Standardized Test Tutor

SCHOLASTIC

GRADE 5

MATH

Practice Tests With Problem-by-Problem Strategies and Tips That Help Students Build Test-Taking Skills and Boost Their Scores

Michael Priestley

Editor: Maria L. Chang
Cover design by Brian LaRossa
Interior design by Creative Pages, Inc.
Illustrations by Creative Pages, Inc.

ISBN-13: 978-0-545-09607-2
ISBN-10: 0-545-09607-3
Copyright © 2009 by Michael Priestley

1   2   3   4   5   6   7   8   9   10   40   15 14 13 12 11 10 09

Contents

Welcome to *Test Tutor* 4

    Test 1 8

    Test 2 28

    Test 3 43

    Answer Sheet 61

Answer Keys

    Test 1 63

    Test 2 73

    Test 3 83

Scoring Charts

    Student Scoring Chart 93

    Classroom Scoring Chart 94

Welcome to *Test Tutor*!

Students in schools today take a lot of tests, especially in reading and math. Some students naturally perform well on tests, and some do not. But just about everyone can get better at taking tests by learning more about what's on the test and how to answer the questions. How many students do you know who could benefit from working with a tutor? How many would love to have someone sit beside them and help them work their way through the tests they have to take?

That's where *Test Tutor* comes in. The main purpose of *Test Tutor* is to help students learn what they need to know in order to do better on tests. Along the way, *Test Tutor* will help students feel more confident as they come to understand the content and learn some of the secrets of success for multiple-choice tests.

The Test Tutor series includes books for reading and books for math in a range of grades. Each *Test Tutor* book in mathematics has three full-length practice tests designed specifically to resemble the state tests that students take each year. The math skills measured on these practice tests have been selected from an analysis of the skills tested in ten major states, and the questions have been written to match the multiple-choice format used in most states.

The most important feature of this book is the friendly Test Tutor. He will help students work through the tests and achieve the kind of success they are looking for. This program is designed so students may work through the tests independently by reading the Test Tutor's helpful hints. Or you may work with the student as a tutor yourself, helping him or her understand each problem and test-taking strategy along the way. You can do this most effectively by following the Test Tutor's guidelines included in the pages of this book.

Three Different Tests

There are three practice tests in this book: Test 1, Test 2, and Test 3. Each test has 50 multiple-choice items with four answer choices (A, B, C, D). All three tests measure the same skills in almost the same order, but they provide different levels of tutoring help.

Test 1 provides step-by-step guidance to help students work through each problem, as in the sample on the next page. The tips in Test 1 are detailed and thorough, and they are written specifically for each math item to help students figure out how to solve the problem.

Sample 1

A school orders 8 cases of juice drinks. Each case has 24 juice drinks. How many juice drinks does the school order in all?

Ⓐ 3

Ⓑ 16

Ⓒ 32

Ⓓ 192

> To find the total number of juice drinks the school orders, find how many drinks there are in 8 cases. Multiply the number of cases times the number of drinks in each case.

Test 2 provides a test-taking tip for each item, as in the sample below, but the tips are less detailed than in Test 1. They help guide the student toward the solution to each problem without giving away too much. Students must take a little more initiative.

Sample 2

At the Indianapolis 500 auto race in 2005, the winner's average speed was 157.6 miles per hour. In 1955, the winner's average speed was 128.2 miles per hour. How much faster was the winner's average speed in 2005?

Ⓐ 29.4 mph

Ⓑ 31.4 mph

Ⓒ 142.9 mph

Ⓓ 285.8 mph

> Look for key words to help you understand what the question is asking.

Test 3 does not provide test-taking tips. It assesses the progress students have made. After working through Tests 1 and 2 with the help of the Test Tutor, students should be more than ready to score well on Test 3 without too much assistance. Success on this test will help students feel confident and prepared for taking real tests.

Other Helpful Features

In addition to the tests, this book provides some other helpful features. First, on page 61, you will find an **answer sheet**. When students take the tests, they may mark their answers by filling in bubbles on the test pages. Or, they may mark their answers on a copy of the answer sheet instead, as they will be required to do in most standardized tests. You may want to have students mark their answers on the test pages for Test 1 and then use an answer sheet for Tests 2 and 3 to help them get used to filling in bubbles.

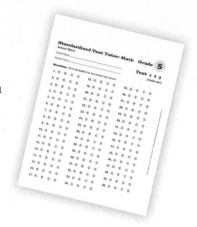

Second, beginning on page 63, you will find a detailed **answer key** for each test. The answer key lists the correct response and explains how to solve the problem. It also identifies the skill tested by each question, as in the sample below.

Answer Key for Sample 1

Correct response: **D**
(*Add, subtract, multiply, and divide whole numbers*)
The key to solving this problem is choosing the correct operation. To find how many juice drinks the school orders, multiply the number of cases (8) by the number of drinks in each case (24): $8 \times 24 = 192$ juice drinks.

Incorrect choices:

A is the result of dividing the number of drinks by the number of cases ($24 \div 8 = 3$) instead of multiplying.

B is the result of subtracting $24 - 8$ instead of multiplying.

C is the result of adding $24 + 8$ instead of multiplying.

As the sample indicates, this question measures the student's ability to add, subtract, multiply, and divide whole numbers. This information can help you determine which skills the student has mastered and which ones still cause difficulty.

Finally, the answer key explains why each incorrect answer choice, or "distractor," is incorrect. This explanation can help reveal what error the student might have made. For example, one distractor in an addition problem might be the result of subtracting two numbers instead of adding them together. Knowing this could help the student understand that he or she used the wrong operation.

At the back of this book, you will find two scoring charts. The **Student Scoring Chart** can help you keep track of each student's scores on all three tests and in different subtests, such as "Number and Number Sense" or "Measurement and Geometry." The **Classroom Scoring Chart** can be used to record the scores for all students on all three tests, illustrating the progress they have made from Test 1 to Test 3. Ideally, students should score higher on each test as they go through them. However, keep in mind that students get a lot of tutoring help on Test 1, some help on Test 2, and no help on Test 3. So if a student's scores on all three tests are fairly similar, that could still be a very positive sign that the student is better able to solve problems independently and will achieve even greater success on future tests.

Directions: Read each question. Look at the Test Tutor's tip for answering the question. Then find the answer. You may do your work on this page or on scrap paper. Mark your answer by filling in the bubble.

1. The table below lists the populations of four states.

State Populations	
Georgia	9,072,576
New Jersey	8,683,242
North Carolina	8,856,505
Virginia	7,567,465

Which state has the greatest population?

Ⓐ Georgia

Ⓑ New Jersey

Ⓒ North Carolina

Ⓓ Virginia

To compare and order these numbers, line them up by place value. The first digit in each of these numbers represents millions.

2. Kate compared the sale prices of bike helmets at four stores. The table below shows the original price and the percent of discount during the sale at each store.

Bike Helmets		
Store	Original Price	Discount Percent
Gerry's Bike Shop	$40	10%
Bikeway	$60	50%
Wheelworks	$50	20%
Big Al's Bikes	$45	10%

Which store has the lowest sale price for a bike helmet?

Ⓐ Gerry's Bike Shop

Ⓑ Bikeway

Ⓒ Wheelworks

Ⓓ Big Al's Bikes

To compare the sale prices, first apply the discount to the original price of each item. For example, the sale price at Gerry's Bike Shop is $40 minus a discount of 10%.

Standardized Test Tutor: Math (Grade 5) © 2009 by Michael Priestley, Scholastic Teaching Resources

Standardized Test Tutor: Math (Grade 5) © 2009 by Michael Priestley, Scholastic Teaching Resources

3. The population of Kansas is about 2,764,075. Which digit in this number represents the thousands place?

Ⓐ 0

Ⓑ 2

Ⓒ 4

Ⓓ 6

> To find place value, read the number 2,764,075 aloud and write it in expanded form. For example, the digit 5 stands for 5 ones.

4. The table below shows the coldest temperatures in four states.

Coldest Temperatures	
State	**Temperature (°F)**
Wisconsin	−55
Wyoming	−66
Utah	−69
Indiana	−36

Which state has the lowest temperature?

Ⓐ Wisconsin

Ⓑ Wyoming

Ⓒ Utah

Ⓓ Indiana

> Remember that greater negative integers have lower value. For example, on a number line, −10 has a lower value than −1.

5. Which group has *all* prime numbers?

Ⓐ 7, 11, 17, 19

Ⓑ 5, 29, 33, 45

Ⓒ 3, 8, 12, 17

Ⓓ 11, 16, 19, 35

> Remember that a prime number has only two factors: 1 and itself. The number 7 is a prime number, but 6 is not.

6. The United States has 14,858 airports. What is this number in expanded form?

Ⓐ 14,000 + 850 + 8

Ⓑ 14,000 + 800 + 50 + 8

Ⓒ 10,000 + 4,800 + 50 + 8

Ⓓ 10,000 + 4,000 + 800 + 50 + 8

> To write a number in expanded form, write the value of each digit. For example, in the number 2,950, the 9 represents 900.

7. In a basketball game, Marcus made 8 of 12 free throws. Which ratio describes Marcus's free throws?

Ⓐ $\frac{12}{8}$

Ⓑ $\frac{2}{3}$

Ⓒ $\frac{3}{4}$

Ⓓ $\frac{8}{20}$

> To find the ratio, compare the number of free throws Marcus made to the total number of free throws, and then simplify.

8. A soccer league for ages 12 and younger has 4 teams of 18 players each. If 20 of these players are 12 years old, how many players are younger than 12?

Ⓐ 20

Ⓑ 52

Ⓒ 56

Ⓓ 60

> Find the total number of players by multiplying the number of teams by the number of players on each team. Then subtract the number of 12-year-olds.

9. The hawk moth flies at a speed of $33\frac{3}{10}$ miles per hour. The dragonfly can fly $17\frac{4}{5}$ miles per hour. How much faster does the hawk moth fly than the dragonfly?

Ⓐ $51\frac{1}{10}$ mph

Ⓑ $16\frac{1}{2}$ mph

Ⓒ $15\frac{4}{5}$ mph

Ⓓ $15\frac{1}{2}$ mph

> To find the difference in speed, subtract the two mixed numbers. Use a common denominator of 10.

Standardized Test Tutor: Math (Grade 5) © 2009 by Michael Priestley, Scholastic Teaching Resources

Standardized Test Tutor: Math (Grade 5) © 2009 by Michael Priestley, Scholastic Teaching Resources

Test Tutor says:

10. The table below shows the number of pounds of cheese produced each week by the Clover Dairy.

Week	Cheese Produced (in pounds)
1	46.7
2	41.4
3	40.2
4	40.1

How many pounds of cheese did the dairy produce in 4 weeks?

Ⓐ 42.1

Ⓑ 167.4

Ⓒ 168.4

Ⓓ 186.8

> To find the total amount of cheese produced, add the amounts for Weeks 1 to 4.

11. Four friends played miniature golf at a course that has a par score of 54. A player who shoots par for the course would have a score of 54. The table below shows each player's scores above ($+$) and below ($-$) par in two rounds.

Meadow Miniature Golf				
Player	Chris	Melinda	Joanna	Linda
Round 1 Score	0	+3	−5	−4
Round 2 Score	0	−2	−1	0

Which player had a combined score of 104 after two rounds?

Ⓐ Chris

Ⓑ Melinda

Ⓒ Joanna

Ⓓ Linda

> Note that each player's score is recorded as the number of strokes above or below 54. A player who scores +1 has a score of $54 + 1 = 55$.

12. The school cafeteria has 28 round tables that seat 11 students each. School workers added 9 more round tables with chairs. Which is the best estimate of the total number of students that will be able to sit at the tables in the cafeteria?

Ⓐ 250

Ⓑ 300

Ⓒ 370

Ⓓ 400

> Round each number to the nearest tens place. For example, 28 would be 30.

13. What is the value of the expression below?

$(15 \div 3) - 2 + (4 \times 5)$

Ⓐ 23

Ⓑ 25

Ⓒ 27

Ⓓ 30

> Follow the order of operations (PEMDAS) to solve this problem: parentheses first, then multiplication and division from left to right, then addition and subtraction from left to right.

14. Which expression is the inverse operation of $320 \div 40$?

Ⓐ $320 + 40$

Ⓑ $40 \div 320$

Ⓒ 40×8

Ⓓ $320 - 40$

> The inverse is the opposite operation.

15. Dora spent a total of $50 to buy paper products, balloons, and flowers for a party. She spent $15.50 on paper products and $12 on balloons. Which number sentence shows how much Dora spent on flowers?

Ⓐ $\$50 - (\$15.50 + \$12) = \$22.50$

Ⓑ $\$15.50 + \$12 = \$27.50$

Ⓒ $\$50 - (\$15.50 - \$12) = \$46.50$

Ⓓ $\$50 + (\$15.50 + \$12) = \$77.50$

> Write a number sentence to solve this problem. Then look at the answers to see if your number sentence is there.

Standardized Test Tutor: Math (Grade 5) © 2009 by Michael Priestley, Scholastic Teaching Resources

16. Jake collected a total of 110 baseball cards over three months. He collected 45 cards in the first month and 27 cards in the second month. How many cards did Jake collect in the third month?

Ⓐ 37

Ⓑ 38

Ⓒ 65

Ⓓ 83

To find the number of cards Jake collected in the third month, subtract the number of cards he collected in the first two months from the total number collected.

17. If $c = 7$, then what is the value of $\frac{4 + 8c}{3}$?

Ⓐ 6.33

Ⓑ 18.67

Ⓒ 20

Ⓓ 60

To find the value of this expression, substitute 7 for c, and then simplify. Make sure to follow the correct order of operations.

18. Sam is 5 feet 2 inches tall. His younger brother, Andy, is 4 inches shorter. How tall is Andy?

Ⓐ 56 inches

Ⓑ 58 inches

Ⓒ 60 inches

Ⓓ 64 inches

To find Andy's height, convert Sam's height to inches and then subtract.

19. Which is the most reasonable measurement for the length of a house key?

Ⓐ 5 centimeters

Ⓑ 5 kilograms

Ⓒ 5 meters

Ⓓ 5 millimeters

Picture a house key in your mind and compare it to the length of a ruler.

20. Fred has a rectangular vegetable garden that is 8 feet long and 6 feet wide. What is the perimeter of the garden?

Ⓐ 14 feet

Ⓑ 28 feet

Ⓒ 30 feet

Ⓓ 48 feet

Remember there are two lengths and two widths in a rectangle.

21. Maura's rectangular bedroom has a perimeter of 36 feet. The width of the room is 8 feet. What is the length of the room?

Ⓐ 20 feet

Ⓑ 16 feet

Ⓒ 10 feet

Ⓓ 8 feet

In a rectangular room, the opposite walls are the same length, so two of the walls in this room measure 8 feet. Subtract these walls from the perimeter to find the lengths of the remaining walls.

22. The parts of a circle are shown below.

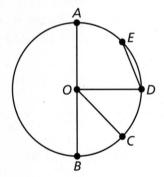

Which line segment identifies the diameter?

Ⓐ $\overline{OC}$

Ⓑ $\overline{OD}$

Ⓒ $\overline{DE}$

Ⓓ $\overline{AB}$

Remember that the diameter is the longest segment in a circle.

Standardized Test Tutor: Math (Grade 5) © 2009 by Michael Priestley, Scholastic Teaching Resources

23. Vera drew a triangle with no equal sides and one angle that is greater than 90°.

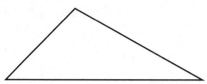

What kind of triangle did she draw?

Ⓐ   acute

Ⓑ   obtuse

Ⓒ   right

Ⓓ   equilateral

> Remember that a right angle measures 90° and an acute angle measures less than that.

24. What is the name of the figure shown below?

Ⓐ   octagon

Ⓑ   rhombus

Ⓒ   hexagon

Ⓓ   pentagon

> Look at the names of the figures. Find the name with a prefix that means "eight."

25. Which figure appears to be similar to the triangle shown below?

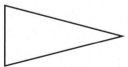

> Keep in mind that similar figures have the same shape but not the same size.

Ⓐ   Ⓒ

Ⓑ    Ⓓ

26. What is the name of the solid figure shown below?

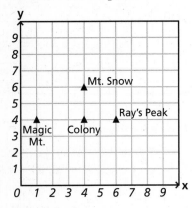

ACME PAINT

1 qt

Ⓐ cone

Ⓑ pyramid

Ⓒ cylinder

Ⓓ prism

Note that the top and bottom of the paint can are round.

27. The coordinate plane shows the location of four mountains.

Which coordinates best describe the location of Mt. Snow?

Ⓐ (1, 4)

Ⓑ (4, 4)

Ⓒ (4, 6)

Ⓓ (6, 4)

To find a location on a coordinate plane, start at 0 and count the number of units left or right and then the number of units up or down.

28. The coordinate plane shows the location of four camping huts in a state park.

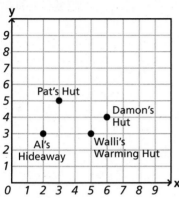

Which camping hut is at (3, 5)?

Ⓐ Walli's Warming Hut

Ⓑ Pat's Hut

Ⓒ Damon's Hut

Ⓓ Al's Hideaway

29. Which of these is a line segment?

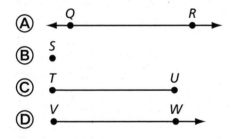

Test Tutor says:

30. Which letter has one line of symmetry?

Ⓐ T

Ⓑ S

Ⓒ P

Ⓓ F

> Use your pencil to draw a line down the middle of each figure. A figure that has symmetry will have the same shape on both sides of the line.

31. Which pair of figures shows a rotation?

Ⓐ

Ⓑ

Ⓒ

Ⓓ

> A rotation is a turn. Look for the figure that turns.

32. Lynne's patio is a square and has a perimeter of 32 feet. What is the area of her patio?

Ⓐ   8 square feet

Ⓑ   16 square feet

Ⓒ   32 square feet

Ⓓ   64 square feet

> The patio is a square, so it has 4 sides of equal length. Divide the perimeter by 4 to find the length of each side. Then multiply to find the area.

Standardized Test Tutor: Math (Grade 5) © 2009 by Michael Priestley, Scholastic Teaching Resources

33. Sara gets up at 9:30 A.M. on Saturday. She spends 40 minutes eating and getting dressed and then practices the piano for 45 minutes. Then she does her homework for 1 hour 20 minutes before she leaves for swim practice. What time does Sara leave for swim practice?

Ⓐ 12:05 P.M.

Ⓑ 12:15 P.M.

Ⓒ 12:20 P.M.

Ⓓ 12:25 P.M.

Write down each step in this problem. Keep in mind that Sara gets up at 9:30 A.M.

34. The height of the Empire State Building in New York City is 1,250 feet. Doug is making a model of the building at a scale of 1 inch = 250 feet. How tall should the model be?

Ⓐ 5 inches

Ⓑ 4 inches

Ⓒ 3 inches

Ⓓ 2 inches

A scale represents a ratio. To find the height of the model, write a proportion using two ratios.

35. The graph below shows the speed some animals can run.

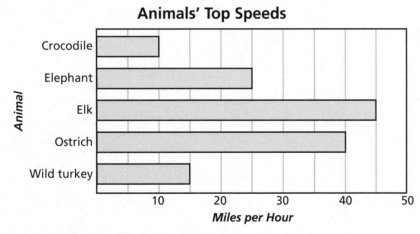

Animals' Top Speeds

Which animal can run 3 times as fast as the wild turkey?

Ⓐ crocodile

Ⓑ elephant

Ⓒ elk

Ⓓ ostrich

Look at the graph to find the speed of the wild turkey. Then multiply the turkey's speed by 3 and find the bar that represents that number.

36. Deena asked 100 students to name their favorite kind of juice. The results are shown in the graph below.

Favorite Kinds of Juice

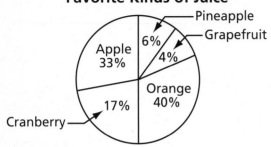

What percentage of students did not choose orange or grapefruit juice?

Ⓐ   44%

Ⓑ   46%

Ⓒ   56%

Ⓓ   60%

> Look at the graph and add the percentages of students who chose a juice that was *not* orange or grapefruit.

Standardized Test Tutor: Math (Grade 5) © 2009 by Michael Priestley, Scholastic Teaching Resources

37. The list below shows the ages of recent U.S. presidents when
they took office.

54	46	64	69	52	61
56	55	43	62	60	51

Which stem-and-leaf plot shows the same information?

Ⓐ

Stem	Leaf
4	1 2
5	2 3 4 4 9
6	0 1 2 5 6

Ⓑ

Stem	Leaf
4	3 6
5	1 2 4 5 6
6	0 1 2 4 9

Ⓒ

Stem	Leaf
4	0 1 2 4 9
5	1 2 4 5 6
6	3 6

Ⓓ

Stem	Leaf
4	2
5	5
6	5

To read a stem-and-leaf
plot, join each "stem" on
the left with each "leaf" on
the right. All of the numbers
in the 50s, for example, are
displayed with the 5 as a
stem and the second digit
in each number as a leaf.

38. The chart below shows the choices for lunch at the Martin School cafeteria on Wednesday.

Wednesday Lunch at Martin School		
Sandwich	**Fruit**	**Drink**
grilled cheese	fruit cocktail	milk
tuna salad	watermelon	bottled water
turkey club		

Which tree diagram shows all the possible combinations of one sandwich, one fruit, and one drink that students may choose?

Look for the tree diagram that shows all possible combinations. Remember that every combination must have a sandwich, a fruit, and a drink.

Ⓐ

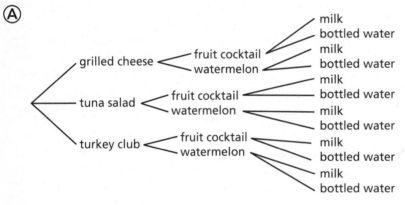

Ⓑ

Ⓒ

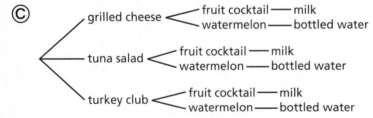

Ⓓ

Standardized Test Tutor: Math (Grade 5) © 2009 by Michael Priestley, Scholastic Teaching Resources

39. Alex writes the letters of the word shown below on index cards and places them in a bag. Then he chooses one card at random.

K A N G A R O O

What is the probability that Alex will choose a vowel?

Ⓐ $\frac{1}{8}$

Ⓑ $\frac{1}{4}$

Ⓒ $\frac{1}{2}$

Ⓓ $\frac{5}{8}$

There are 8 letters in the word *kangaroo*, so there will be 8 index cards. Count the number of letters that are vowels.

40. Ken places 9 red marbles, 6 blue marbles, 4 green marbles, and 1 yellow marble in a bag. If he chooses one marble from the bag without looking, which color marble is he least likely to pick?

Ⓐ red

Ⓑ blue

Ⓒ green

Ⓓ yellow

Find the total number of marbles in the bag. Then compare the number of marbles of each color in relation to the total.

Use the table to answer questions 41–42. It shows the highest temperatures ever recorded in seven states.

Highest Temperatures	
State	**Temperature (°F)**
Arizona	128
California	134
Florida	109
Nevada	125
Rhode Island	104
Texas	120
Wyoming	115

41. What is the range of these temperatures?

Ⓐ 30°

Ⓑ 119°

Ⓒ 120°

Ⓓ 134°

> Note that the range is the difference between the highest and the lowest numbers in a set of data.

42. What is the median temperature shown?

Ⓐ 104°

Ⓑ 115°

Ⓒ 120°

Ⓓ 134°

> To find the median temperature, list the temperatures in order and find the number in the middle of the list.

Standardized Test Tutor: Math (Grade 5) © 2009 by Michael Priestley, Scholastic Teaching Resources

43. Martha listed the number of U.S. Representatives in each state near her home state of Colorado. The table below shows the results.

State	U.S. Representatives
Colorado	7
Wyoming	1
Nebraska	3
Kansas	4
Oklahoma	5
New Mexico	3
Arizona	8

What type of graph would best represent these data?

Ⓐ bar graph

Ⓑ circle graph

Ⓒ stem-and-leaf plot

Ⓓ line graph

> Look for the type of graph that displays data as separate values.

44. The figure below shows a repeating pattern.

If the pattern continues, what will be the 11th shape?

Ⓐ

Ⓑ

Ⓒ

Ⓓ

> Draw the figures that would continue the pattern until you get to the 11th shape.

45. Sharon collected a number of bottles (b) for a bottle drive. Kayla collected 32 more than twice the number of bottles Sharon collected. Which expression describes the number of bottles (b) Kayla collected?

Ⓐ $32 - 2b$

Ⓑ $2 + (32 + b)$

Ⓒ $32 \times 2b$

Ⓓ $32 + 2b$

> Think of the number of bottles Kayla collected in relation to the number Sharon collected. Write an expression for this number and see if it matches any of the answer choices.

46. At a garden shop, Mario buys 3 cornflower plants for $9.95 each and 3 black-eyed Susans for m each. Which expression represents the total amount Mario spent?

Ⓐ $3 \times (9.95 + m)$

Ⓑ $3 + (9.95 + m)$

Ⓒ $(9.95 + m) \div 3$

Ⓓ $(3 \times 9.95) \times (3 \times m)$

> Write an expression based on the information in the problem. Then look for an equivalent expression in the answer choices.

47. Which situation can best be represented by the number sentence $54 - n = 41$?

Ⓐ Jack had 41 baseball cards. He bought 13 new cards and now has 54.

Ⓑ Beth made 54 bracelets and gave some away. She now has 41 bracelets.

Ⓒ Sue has 54 shells, Meg has 13 shells, and Rita has 41 shells.

Ⓓ Mark divided 54 sandwiches among his friends. Now he has 41 left.

> Read each situation carefully to figure out which number is unknown. That unknown number is n.

Standardized Test Tutor: Math (Grade 5) © 2009 by Michael Priestley, Scholastic Teaching Resources

Test Tutor says:

48. Tyler and Cameron plan to rent bikes and helmets for a day of bike-riding. Renting a bike helmet for the day costs $8. The table below shows the cost of renting a bike by the hour.

Bike Rentals

Number of Hours	1	2	3	4	5	6
Cost	$8	$11	$14	$17		

Look for the pattern in the table by determining how the cost changes for each additional hour.

How much will it cost each boy to rent a bike helmet and a bike for 8 hours?

Ⓐ $25   Ⓒ $33

Ⓑ $29   Ⓓ $37

49. If $5x + 7 = 42$, then what is the value of x?

Ⓐ 8   Ⓒ 6

Ⓑ 7   Ⓓ 5

To solve the equation, you must first get x on one side of the equation by itself. You must add or subtract before you multiply or divide.

50. Janice signed up for a 12-week exercise class. The table below shows the number of push-ups she does daily each week.

Week	Number of Push-Ups
1	6
2	10
3	14
4	18

At this rate, how many push-ups will she do daily in week 9?

Ⓐ 22

Ⓑ 30

Ⓒ 34

Ⓓ 38

Look at the table carefully. How does the number of push-ups change from one week to the next?

End of Test 1   **STOP**

Directions: Read each question. Look at the Test Tutor's tip for answering the question. Then find the answer. You may do your work on this page or on scrap paper. Mark your answer by filling in the correct bubble.

1. Rick is making tomato sauce. He adds $\frac{1}{2}$ teaspoon oregano, $\frac{1}{8}$ teaspoon red pepper, $\frac{1}{4}$ teaspoon basil, and 1 teaspoon sugar. For which ingredient does he use the *least* amount?

   Ⓐ oregano

   Ⓑ red pepper

   Ⓒ basil

   Ⓓ sugar

To compare fractions with different denominators, change them to a common denominator.

2. A total of 600 students goes to the Taft School. The table below shows the percentage of students who attended school each week in January.

Attendance at Taft School	
Week 1	84%
Week 2	88%
Week 3	80%
Week 4	92%

   In which week did fewer than 500 students attend school?

   Ⓐ Week 1

   Ⓑ Week 2

   Ⓒ Week 3

   Ⓓ Week 4

Rewrite percentages as their decimal equivalents.

3. The United States has 3,981,546 miles of roads. Which digit in this number represents the hundred thousands place?

   Ⓐ 9

   Ⓑ 8

   Ⓒ 5

   Ⓓ 3

Write the number in expanded form to find place values.

Standardized Test Tutor: Math (Grade 5) © 2009 by Michael Priestley, Scholastic Teaching Resources

Test Tutor says:

4. The table below shows the low elevations of some famous places.

Place	Elevation (in feet below sea level)
Lake Assai, Africa	−512
Lake Eyre, Australia	−52
Caspian Sea, Russia	−92
Death Valley, California	−282
Valdes Peninsula, Argentina	−132

Which elevation is closest to sea level (0 feet)?

Ⓐ −512   Ⓒ −92

Ⓑ −132   Ⓓ −52

> Which number comes closest to 0 on a number line?

5. Which is a composite number?

Ⓐ 13   Ⓒ 17

Ⓑ 15   Ⓓ 23

> Look for the number that has more than two factors.

6. A park in Greenland has an area of 375,291 square miles. What is this number in expanded form?

Ⓐ 370,000 + 70,000 + 5,000 + 290 + 1

Ⓑ 370,000 + 50,000 + 200 + 90 + 1

Ⓒ 300,000 + 50,000 + 200 + 90 + 1

Ⓓ 300,000 + 70,000 + 5,000 + 200 + 90 + 1

> Cross out the answer choices that do not show the value of every digit.

7. In one season, the Seahawks won 10 games and lost 6 games. What was the team's ratio of wins to games played?

Ⓐ 10:16

Ⓑ 4:16

Ⓒ 10:10

Ⓓ 10:6

> Add the number of wins and losses to find the total number of games.

8. The sum of the ages of Adrian, Manny, and Kyle is 41. Adrian is 14 years old and Manny is 3 years older than Kyle. How old is Kyle?

Ⓐ 24

Ⓑ 17

Ⓒ 15

Ⓓ 12

> Write a number sentence to find Kyle's age.

9. The Senate has 100 members and the House of Representatives has 435 members. To pass an amendment to the Constitution, a $\frac{2}{3}$ majority vote is needed in both the Senate and the House. About how many votes all together will be needed to pass an amendment?

Ⓐ 535

Ⓑ 356

Ⓒ 290

Ⓓ 178

> Look for key words to help you decide what the question is asking and what kind of computation you need to do.

10. The table below shows the weights of three whales.

Whale	Weight (in tons)
Fin whale	49.6
Right whale	41.1
Sperm whale	39.7

What is the total weight of these three whales?

Ⓐ 80.8 tons

Ⓑ 90.7 tons

Ⓒ 130.4 tons

Ⓓ 140.4 tons

> What operation do you need to use to find the total weight?

Standardized Test Tutor: Math (Grade 5) © 2009 by Michael Priestley, Scholastic Teaching Resources

11. The table below shows the temperatures in some of the world's coldest places.

Place	Temperature (in °C)
Eureka, Canada	−19.7
Ostrov Bolshoy, Russia	−14.7
Point Barrow, Alaska	−12.1
Resolute, Canada	−26

Which temperature is 7.6°C colder than Point Barrow, Alaska?

Ⓐ −4.5°C

Ⓑ −14.7°C

Ⓒ −19.7°C

Ⓓ −26°C

> Remember the temperature farthest away from zero is the coldest.

12. The table below shows the lengths of four tunnels in Japan.

Tunnel	Length (in kilometers)
Dai-Shimizu	22.2
Tokyo Bay Aqualine	9.5
Seikan	53.8
Shin-Kanmon	18.8

Which is the closest estimate of the total length of all four tunnels?

Ⓐ 90 km

Ⓑ 100 km

Ⓒ 120 km

Ⓓ 150 km

> Round the numbers to the nearest whole number in order to estimate.

13. What is the value of the expression below?

$(36 \div 3) + 25 - (5 \times 7)$

Ⓐ 224   Ⓒ 32

Ⓑ 72   Ⓓ 2

> Remember to complete the operations in parentheses first.

Test Tutor says:

14. Which expression is the inverse operation of $45 - 36$?

Ⓐ $36 + 9$

Ⓑ 5×9

Ⓒ $36 + 45$

Ⓓ 45×36

> What is the inverse of subtraction?

15. In a class election, Mark received $\frac{3}{5}$ of the votes in his class of 30 students. How many students voted for other candidates?

Ⓐ 10

Ⓑ 12

Ⓒ 15

Ⓓ 18

> Note that you are looking for the number of students who did *not* vote for Mark.

16. Carly works at a local dry cleaner 12 hours each week. She makes $8.25 an hour. How much does she earn in 4 weeks?

Ⓐ $99

Ⓑ $198

Ⓒ $396

Ⓓ $432

> Write a number sentence to include each step in this problem.

17. If $b = 7$, then what is the value of $4b + (6 + 7)$?

Ⓐ 24

Ⓑ 28

Ⓒ 41

Ⓓ 45

> Substitute the value of b in the expression and simplify.

18. Julie buys 2 liters of orange juice. She drinks 250 milliliters of juice for lunch. How much orange juice is left?

Ⓐ 1,750 milliliters

Ⓑ 1,250 milliliters

Ⓒ 750 milliliters

Ⓓ 250 milliliters

> Remember that 1 liter = 1,000 milliliters.

19. Which is the most appropriate unit for measuring the length of a football field?

Ⓐ centimeters

Ⓑ kilometers

Ⓒ milliliters

Ⓓ meters

Look for the unit of measure that is closest to a yard.

20. Hank has a rectangular yard for his dog that measures 16 feet by 10 feet. He wants to double the area of the yard. What could be the dimensions of the new yard?

Ⓐ 8 ft × 10 ft

Ⓑ 8 ft × 20 ft

Ⓒ 32 ft × 20 ft

Ⓓ 10 ft × 32 ft

What is the area of the yard now? Which answer choice is double that?

21. The perimeter of a rectangular pool is 40 feet. The length of the pool is 12 feet. What is the area of the pool?

Ⓐ 8 square feet

Ⓑ 20 square feet

Ⓒ 96 square feet

Ⓓ 192 square feet

Remember the area equals width × length.

22. Which term describes *AO* in the figure below?

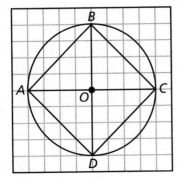

Ⓐ chord

Ⓑ diameter

Ⓒ radius

Ⓓ ray

What is the name of the segment that has one endpoint on the circle and one endpoint at the center?

Test Tutor says:

23. Jerry drew a triangle with sides that measure 8 inches, 10 inches, and 7 inches. What kind of triangle did he draw?

Ⓐ isosceles

Ⓑ scalene

Ⓒ equilateral

Ⓓ right

Draw your own picture if it will help you solve the problem.

24. Which figure has six sides and six angles?

 Ⓐ

 Ⓒ

 Ⓑ

 Ⓓ

Count the number of sides and angles in each figure.

25. In which pair are the figures congruent?

 Ⓐ

 Ⓒ

 Ⓑ

 Ⓓ

Remember that congruent figures are identical.

26. What is the name of the figure shown below?

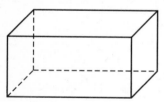

Ⓐ rectangular prism

Ⓑ pyramid

Ⓒ cylinder

Ⓓ cone

Cross out every answer that you know is wrong. If more than one is left, make your best guess.

Standardized Test Tutor: Math (Grade 5) © 2009 by Michael Priestley, Scholastic Teaching Resources

27. Saul made a treasure map for his friends, as shown below.

Saul's Treasure Map

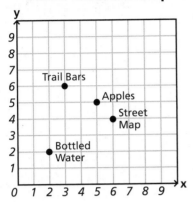

Which coordinates best describe the location of the trail bars?

- (A) (2, 2)
- (B) (3, 6)
- (C) (5, 5)
- (D) (6, 4)

Circle the trail bars on the grid and then look at the answer choices.

28. Hannah made a map of the local zoo, as shown below.

Zoo Map

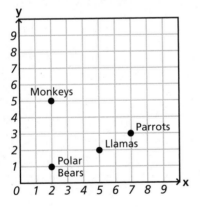

Which animals are located at (2, 5)?

- (A) polar bears
- (B) parrots
- (C) llamas
- (D) monkeys

Remember that the first number in an ordered pair is the *x*-coordinate, which goes across the grid.

29. What is the best name for the geometric figure shown below?

Ⓐ a point

Ⓑ a line segment

Ⓒ a ray

Ⓓ a line

30. Which figure has exactly 3 lines of symmetry?

Ⓐ   Ⓒ

Ⓑ   Ⓓ

31. Garrett drew the figure shown below.

Which of these is a reflection of Garrett's figure?

Ⓐ   Ⓒ

Ⓑ   Ⓓ

32. The temperature at noon was about 26°F. By midnight, the temperature was −4°F. How many degrees did the temperature drop?

Ⓐ 22°

Ⓑ 24°

Ⓒ 26°

Ⓓ 30°

Standardized Test Tutor: Math (Grade 5) © 2009 by Michael Priestley, Scholastic Teaching Resources

Test
Tutor
says:

33. An ostrich egg weighs 4 pounds. How many ounces does it weigh?

Ⓐ 96 ounces

Ⓑ 64 ounces

Ⓒ 40 ounces

Ⓓ 32 ounces

How many ounces are in one pound?

34. Julie is making a model of the cruise ship *Queen Mary 2*. The ship is 1,132 feet long and 236 feet wide. Julie's scale is 1 inch = 200 feet. About how long will the model be?

Ⓐ 10 inches

Ⓑ 8 inches

Ⓒ 6 inches

Ⓓ 4 inches

Draw a picture if it will help you solve the problem.

35. The graph below shows average monthly temperatures in San Diego, California.

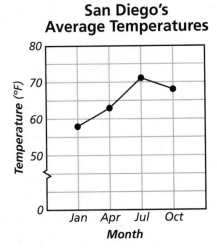

Circle the two points you need to focus on and compare them.

What is the difference in temperatures between January and October?

Ⓐ 3°   Ⓒ 8°

Ⓑ 5°   Ⓓ 10°

Test Tutor says:

36. The graph below shows the populations of five European countries.

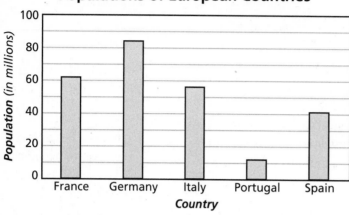

Populations of European Countries

Which country has about $\frac{1}{2}$ the population of Germany?

Ⓐ France   Ⓒ Portugal

Ⓑ Italy   Ⓓ Spain

Find Germany's population, then divide that number in half.

37. The ages of members in the Madison bowling league are shown in the stem-and-leaf plot below.

Ages of League Bowlers

Stem	Leaf
1	0 0 1 1 2 2 2 3
2	0 1 9
3	3 3 4 4 5 5 7 8 9
4	0 0 1 1
5	2

How many bowlers in the league are 20 years old and older?

Ⓐ 14

Ⓑ 16

Ⓒ 17

Ⓓ 25

Note that this question asks how many players are *20 years old and older.*

Standardized Test Tutor: Math (Grade 5) © 2009 by Michael Priestley, Scholastic Teaching Resources

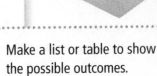

38. Latoya rolls a number cube numbered 1 to 6. If she rolls the cube 12 times, how many times is she likely to roll a 2 or a 3?

Ⓐ 2

Ⓑ 4

Ⓒ 6

Ⓓ 8

39. Zach has a fair spinner with 4 colored sections of equal size: red, blue, yellow, and green. If he spins the spinner 40 times, about how many times will it land on yellow?

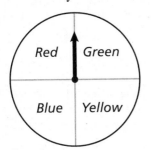

Ⓐ 4

Ⓑ 10

Ⓒ 15

Ⓓ 20

40. A bag contains 9 letter tiles: A, B, C, D, E, F, G, H, I. If Dan selects one letter tile without looking, what is the probability that he will pick a vowel?

Ⓐ $\frac{1}{3}$

Ⓑ $\frac{1}{2}$

Ⓒ $\frac{2}{3}$

Ⓓ $\frac{2}{9}$

Test Tutor says:

Use the table to answer questions 41–42. It shows the numbers of endangered and threatened species in the United States.

Endangered and Threatened Species in the United States	
Mammals	81
Birds	89
Reptiles	37
Amphibians	23
Fishes	139
Clams	70
Snails	75

41. What is the median of the data in the table?

  Ⓐ 81   Ⓒ 73

  Ⓑ 75   Ⓓ 70

List the numbers from the table in order to find the middle number.

42. What is the range of the data?

  Ⓐ 50   Ⓒ 116

  Ⓑ 102   Ⓓ 162

Use the highest and lowest numbers from the table to find the range.

43. Beth read 20 books during the summer. The graph below shows the kinds of books she read.

Beth's Summer Reading

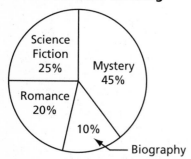

How many mystery books did Beth read?

  Ⓐ 8   Ⓒ 10

  Ⓑ 9   Ⓓ 45

Read the problem carefully so you know what it is asking for.

Standardized Test Tutor: Math (Grade 5) © 2009 by Michael Priestley, Scholastic Teaching Resources

Standardized Test Tutor: Math (Grade 5) © 2009 by Michael Priestley, Scholastic Teaching Resources

44. Jack started a stamp collection with 19 stamps that his uncle gave him. Every two months, Jack added to his collection, as shown in the table below.

Month	Jan	Mar	May	Jul	Sep	Nov
Number of Stamps	19	31	43	55		

If this pattern continues, how many stamps will Jack have the next January?

Ⓐ 79

Ⓑ 85

Ⓒ 90

Ⓓ 91

> Use the numbers you know to find the rule for this table and determine the missing number.

45. What is the value of the expression $248 \div 2a$, when $a = 4$?

Ⓐ 31   Ⓒ 124

Ⓑ 62   Ⓓ 256

> Substitute the value of *a* to simplify this expression.

46. The school bought 24 tickets for the aquarium. Tickets cost $18 for students and $30 for adults. Which expression shows the total cost of tickets for 20 students and 4 adults?

Ⓐ $(18 + 30) \times 24$

Ⓑ $(20 \times 18) \times (4 \times 30)$

Ⓒ $(20 \times 18) + (4 \times 30)$

Ⓓ $(20 \times 18) - (4 \times 30)$

> Write a number sentence to solve this problem and then compare it with the answer choices.

47. Which situation can best be described by the number sentence $7x = 420$?

Ⓐ June gave away 7 books to her friend. Now she has 420 books.

Ⓑ Theo worked 7 hours on Monday, on Wednesday, and on Friday.

Ⓒ Carol had 420 beads that she divided among 7 friends.

Ⓓ On his way to Oregon, Jack drove 420 miles in 7 hours.

> Write a number sentence for each situation to see which one works.

48. Mariel spent $980 during her vacation in Alabama. The graph below shows how she spent her money.

Mariel's Expenses

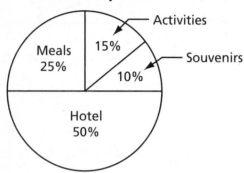

Which statement best describes Mariel's expenses?

(A) She spent most of her money on meals and activities.

(B) She spent one-half of her money on the hotel.

(C) She spent more on meals than she spent on hotel and activities.

(D) She spent less than one-quarter of her money on meals and souvenirs.

> Look at the circle graph carefully to find the information you need.

49. If $72 \div x = 8$, then what is the value of x?

(A) 7   (C) 12

(B) 9   (D) 80

> Try each answer choice in the equation to find the one that works.

50. Gwen draws the pattern shown below.

If the pattern continues, what will be the 15th figure in the pattern?

(A)   (C)

(B)   (D)

> Study the pattern to determine how the figures change. How many figures are in the pattern before it repeats?

End of Test 2   **STOP**

Standardized Test Tutor: Math (Grade 5) © 2009 by Michael Priestley, Scholastic Teaching Resources

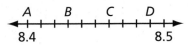

Good Luck!

Directions: Choose the best answer to each question. Mark your answer by filling in the correct bubble.

1. Which point on the number line represents a number that is 0.03 greater than 8.43?

   ```
        A      B      C      D
     ←──┼──┼──┼──┼──┼──┼──┼──┼──┼──→
       8.4                  8.5
   ```

   Ⓐ Point *A*

   Ⓑ Point *B*

   Ⓒ Point *C*

   Ⓓ Point *D*

2. The Rovers soccer team won 12 games and lost 4 games. What percentage of their games did the soccer team lose?

   Ⓐ 25%

   Ⓑ $33\frac{1}{3}$%

   Ⓒ 67%

   Ⓓ 75%

3. A ship dropped its anchor from a height of 5 feet above the water to a depth of −22 feet. How many feet did the anchor drop?

   Ⓐ 13 feet

   Ⓑ 17 feet

   Ⓒ 25 feet

   Ⓓ 27 feet

4. A basket of fruit weighs 732.568 grams. Which digit in this number represents the thousandths place?

   Ⓐ 5

   Ⓑ 6

   Ⓒ 7

   Ⓓ 8

5. Mark has 84 books. He wants to put them into boxes with the same number of books in each box. Which of these could be the number of books in each box?

(A) 16

(B) 14

(C) 10

(D) 8

6. What is the value of 5^3?

(A) 125

(B) 45

(C) 25

(D) 15

7. The Student Council is made up of 5 boys and 7 girls. What is the ratio of the number of girls to the total number of Student Council members?

(A) 5 to 7

(B) 5 to 12

(C) 7 to 12

(D) 12 to 7

8. Hawaii has a total area of 10,931 square miles. New Hampshire has an area of 9,350 square miles. How much larger is Hawaii than New Hampshire?

(A) 581 square miles

(B) 1,581 square miles

(C) 1,621 square miles

(D) 20,281 square miles

Standardized Test Tutor: Math (Grade 5) © 2009 by Michael Priestley, Scholastic Teaching Resources

9. Darrell buys $\frac{3}{4}$ pound of potato salad. He gives $\frac{2}{5}$ of the potato salad to his friend Randy. How much potato salad does he give to Randy?

   Ⓐ $\frac{1}{4}$ pound

   Ⓑ $\frac{3}{10}$ pound

   Ⓒ $\frac{7}{20}$ pound

   Ⓓ $\frac{6}{9}$ pound

10. Camille rides her bike 2.7 miles to school every morning and 2.7 miles home every afternoon. How many miles does she ride in 5 days?

    Ⓐ 10.4 miles

    Ⓑ 13.5 miles

    Ⓒ 27 miles

    Ⓓ 36.45 miles

11. During a football game, the Rams gained 23 yards on the first play. On the second play, they lost 7 yards. On the third play, they gained 5 yards. What is the total number of yards they gained or lost on the three plays?

    Ⓐ −35 yards

    Ⓑ 11 yards

    Ⓒ 21 yards

    Ⓓ 35 yards

12. Ben plans to walk 15 miles a week. He walks $2\frac{1}{4}$ miles on Monday, $3\frac{1}{3}$ miles on Tuesday, and $3\frac{7}{10}$ miles on Wednesday. About how many more miles does he need to walk to reach his goal?

    Ⓐ 9

    Ⓑ 8

    Ⓒ 7

    Ⓓ 6

13. What is the value of the expression below?

$(21 - 6) + 3^2 - (8 \times 2)$

Ⓐ 5

Ⓑ 8

Ⓒ 20

Ⓓ 32

14. Which expression is the inverse operation of 30×20?

Ⓐ 30 + 20

Ⓑ 30 − 20

Ⓒ 30 ÷ 20

Ⓓ 600 ÷ 30

15. Students at the Washington School held a car wash one Saturday to raise money for new computers. They charged $7 for cars and $9 for vans. At the end of the day, students had washed 24 cars and 15 vans. How much money did they earn all together?

Ⓐ $135

Ⓑ $168

Ⓒ $303

Ⓓ $320

Standardized Test Tutor: Math (Grade 5) © 2009 by Michael Priestley, Scholastic Teaching Resources

16. In Mr. Bell's class, 24 students voted on their first choice for a field trip. The graph below shows the results.

Field Trip Choices

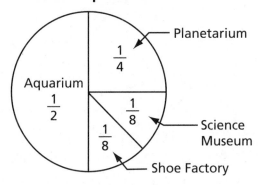

How many students in all chose the aquarium or the science museum?

Ⓐ 18

Ⓑ 15

Ⓒ 12

Ⓓ 9

17. If $y = 8$, then what is the value of $\dfrac{56 - 2y}{4}$?

Ⓐ 10

Ⓑ 16

Ⓒ 40

Ⓓ 72

18. A dog named Max weighs 9 kilograms. About how many pounds does Max weigh? (1 kilogram = 2.2 pounds)

Ⓐ 4

Ⓑ 10

Ⓒ 20

Ⓓ 144

19. Which would be the most appropriate unit for measuring the capacity of a swimming pool?

Ⓐ tons

Ⓑ ounces

Ⓒ quarts

Ⓓ gallons

20. John made a triangular kite. The base of the kite is 20 inches and the height is 12 inches. What is the area of the kite?

Ⓐ 80 square inches

Ⓑ 120 square inches

Ⓒ 160 square inches

Ⓓ 240 square inches

21. The diagram below shows the measurements of Amy's vegetable garden. What is the area of the garden?

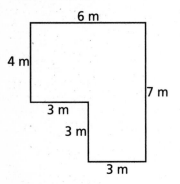

Ⓐ 24 m²

Ⓑ 26 m²

Ⓒ 33 m²

Ⓓ 42 m²

Standardized Test Tutor: Math (Grade 5) © 2009 by Michael Priestley, Scholastic Teaching Resources

22. A round table in the cafeteria has a diameter of 8 feet. Which expression can be used to find the circumference of the table?

(A) π × 4 feet

(B) π × 8 feet

(C) π × 16 feet

(D) π × 64 feet

23. Which figure is an acute triangle?

(A)

(B)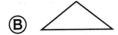

(C)

(D)

24. Which phrase best describes a rhombus?

(A) a quadrilateral with two pairs of parallel sides and four equal sides

(B) a polygon that has three sides and three angles

(C) a quadrilateral with exactly two parallel sides

(D) a polygon that has six angles and six sides of equal length

25. Which two figures appear to be both similar and congruent?

(A)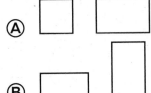

(B)

(C)

(D)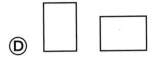

26. What kind of figure is shown below?

- (A) cylinder
- (B) cone
- (C) triangle
- (D) pyramid

27. The coordinate plane represents the floor plan of a school cafeteria.

School Cafeteria

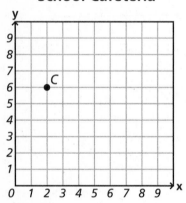

The cashier is located at point *C*. The salad bar is located 4 units to the right and 3 units down from the cashier. What are the coordinates for the salad bar?

- (A) (4, 3)
- (B) (4, 6)
- (C) (6, 3)
- (D) (5, 2)

Standardized Test Tutor: Math (Grade 5) © 2009 by Michael Priestley, Scholastic Teaching Resources

28. Look at the figure below.

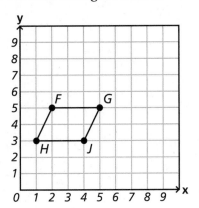

Which point is located at (5, 5)?

Ⓐ F

Ⓑ G

Ⓒ H

Ⓓ J

29. Which figure is a line?

Ⓐ

Ⓑ P ●————————● Q

Ⓒ R ●

Ⓓ S ●————————→ T

30. Which figure has exactly 4 lines of symmetry?

Ⓐ

Ⓑ

Ⓒ

Ⓓ

31. Which pair of figures shows a translation?

Ⓐ

Ⓑ

Ⓒ

Ⓓ

32. The area of a rectangular shed is 96 square feet. The width of the shed is 8 feet. What is the length of the shed?

Ⓐ   8 feet

Ⓑ   12 feet

Ⓒ   20 feet

Ⓓ   40 feet

33. The figure below shows the measurements of a rectangular box.

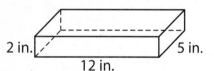

2 in.   5 in.

12 in.

What is the volume of the box?

Ⓐ   240 cubic inches

Ⓑ   120 cubic inches

Ⓒ   84 cubic inches

Ⓓ   60 cubic inches

Standardized Test Tutor: Math (Grade 5) © 2009 by Michael Priestley, Scholastic Teaching Resources

34. On a map, the distance from Bell Harbor to Wayland measures 4 inches. The scale on the map is 1 inch = 4.5 miles. What is the actual distance from Bell Harbor to Wayland?

Ⓐ   18 miles

Ⓑ   16 miles

Ⓒ   13.5 miles

Ⓓ   8.5 miles

35. The chart below shows the length of the subway systems in five major cities.

World's Largest Subway Systems	
City	Lengths (in miles)
New York	231
Paris	126
Tokyo	160
London	244
Moscow	163

How much longer is New York City's subway system than Tokyo's?

Ⓐ   391 miles

Ⓑ   84 miles

Ⓒ   81 miles

Ⓓ   71 miles

36. The graph below shows the average temperatures in Santa Fe, New Mexico.

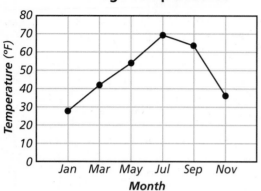

Santa Fe's Average Temperatures

Which month has the lowest average temperature?

(A) January

(B) March

(C) July

(D) November

37. The stem-and-leaf plot lists the heights in inches of tomato plants in Andrew's garden.

Stem	Leaf
2	6 7 9
3	0 1 3 3 5 6 6
4	0 1

How many tomato plants are taller than 29 inches?

(A) 3

(B) 7

(C) 9

(D) 12

38. The chart below shows the menu items customers can choose for the breakfast special at Mel's Diner.

Main Course	Side	Drink
Eggs	Toast	Juice
Omelet	Muffin	Milk
Cereal	Fruit	

Which list shows all of the possible combinations of one main course, one side dish, and one drink?

Ⓐ
```
Eggs ——— Toast ——— Juice
Omelet — Muffin — Milk
Cereal —— Fruit
```

Ⓑ
```
Eggs ——— Toast ——— Juice    Eggs ——— Toast ——— Milk
Eggs ——— Muffin — Juice    Eggs ——— Muffin — Milk
Eggs ——— Fruit ——— Juice    Eggs ——— Fruit ——— Milk

Omelet — Toast ——— Juice    Omelet — Toast ——— Milk
Omelet — Muffin — Juice    Omelet — Muffin — Milk
Omelet — Fruit ——— Juice    Omelet — Fruit ——— Milk

Cereal —— Toast ——— Juice    Cereal —— Toast ——— Milk
Cereal —— Muffin — Juice    Cereal —— Muffin — Milk
Cereal —— Fruit ——— Juice    Cereal —— Fruit ——— Milk
```

Ⓒ
```
Eggs ——— Toast ——— Juice ——— Milk
Eggs ——— Muffin — Juice ——— Milk
Eggs ——— Fruit ——— Juice ——— Milk

Omelet — Toast ——— Juice ——— Milk
Omelet — Muffin — Juice ——— Milk
Omelet — Fruit ——— Juice ——— Milk

Cereal —— Toast ——— Juice ——— Milk
Cereal —— Muffin — Juice ——— Milk
Cereal —— Fruit ——— Juice ——— Milk
```

Ⓓ
```
Eggs ——— Omelet — Cereal
Toast ——— Muffin — Fruit
Juice ——— Milk
```

Questions 39–40: At a carnival, the Duck Pond game has 60 colored ducks floating in a pool. The chart below shows the number of ducks of each color.

Duck Pond					
Colors	Red	Orange	Yellow	Blue	Purple
Number of Ducks	3	10	20	12	15

39. If Martina chooses one duck from the pool without looking, which color is she most likely to get?

Ⓐ red

Ⓑ orange

Ⓒ yellow

Ⓓ purple

40. If Martina chooses one duck from the pool, what is the probability that she will get a blue duck?

Ⓐ 0.12

Ⓑ 0.2

Ⓒ 0.25

Ⓓ 0.33

41. The table below shows the percentages of the different kinds of trash left at a city landfill.

Kind of Trash	**Percent**
Food and yard waste	24%
Metal	7%
Miscellaneous	25%
Paper	25%
Plastic	16%
Rubber and leather	3%

What type of graph would best display these data?

Ⓐ line graph

Ⓑ line plot

Ⓒ circle graph

Ⓓ stem-and-leaf plot

Standardized Test Tutor: Math (Grade 5) © 2009 by Michael Priestley, Scholastic Teaching Resources

Use the chart below to answer questions 42–43. It shows the number of people who visited five popular national parks in the United States.

National Park	Location	Number of Visitors (in millions)
Grand Canyon	Arizona	4.4
Great Smoky Mountains	Tennessee	9.3
Olympic	Washington	2.7
Yellowstone	Wyoming, Montana, Idaho	2.9
Yosemite	California	3.2

42. What is the median number of visitors at these parks?

Ⓐ 2.8 million

Ⓑ 3.2 million

Ⓒ 4.4 million

Ⓓ 6.6 million

43. What is the mean of the data in the table?

Ⓐ 4.5 million

Ⓑ 5.6 million

Ⓒ 6.6 million

Ⓓ 22.5 million

44. What will be the next three numbers in the pattern shown in the table below?

x	y
1	4
3	6
5	8
7	
9	
11	

(A) 10, 13, 15

(B) 10, 12, 14

(C) 9, 11, 13

(D) 8, 9, 10

45. What is the value of the expression $(3x - 4) - 3$, when $x = 5$?

(A) 22

(B) 15

(C) 11

(D) 8

46. Which expression also describes $6 \times (4 + 3)$?

(A) $6 \times (4 \times 3)$

(B) $(6 \times 4) + 3$

(C) $(6 \times 4) + (6 \times 3)$

(D) $(6 + 4) \times 3$

47. Which situation can best be described by the number sentence $63 \div x = 7$?

   Ⓐ Katya had 63 CDs. She gave an equal number of CDs to each of 7 friends.

   Ⓑ Carl has 63 books in his home library. He has read 7 of them so far.

   Ⓒ Sam worked 63 hours in May and another 7 hours in June.

   Ⓓ Myra has $0.63 in her purse. Tony has 7 times that amount in his pocket.

48. The table shows Brenda's bowling scores for five games.

Brenda's Bowling Scores	
Game	Score
1	89
2	96
3	103
4	110
5	117

Which sentence best describes the pattern in these scores?

   Ⓐ Brenda's scores decreased steadily from Game 1 to Game 5.

   Ⓑ Brenda's scores increased by 7 points in each game.

   Ⓒ Brenda nearly doubled her score from Game 1 to Game 5.

   Ⓓ Brenda improved her scores by more than 10 percent each game.

49. If $3x = 159$, then what is the value of x?

   Ⓐ 50

   Ⓑ 53

   Ⓒ 55

   Ⓓ 477

50. The table below shows Kevin's training schedule for jogging.

Kevin's Training Schedule	
Week	Number of Miles
1	2
2	5
3	8
4	11

If Kevin continues with his training pattern, how many miles will he run during Week 6?

Ⓐ 12

Ⓑ 14

Ⓒ 15

Ⓓ 17

End of Test 3   **STOP**

Standardized Test Tutor: Math Grade

Answer Sheet

Student Name _____

Teacher Name _____

Test 1 2 3

(circle one)

Directions: Fill in the bubble for the answer you choose.

1. Ⓐ Ⓑ Ⓒ Ⓓ   18. Ⓐ Ⓑ Ⓒ Ⓓ   35. Ⓐ Ⓑ Ⓒ Ⓓ
2. Ⓐ Ⓑ Ⓒ Ⓓ   19. Ⓐ Ⓑ Ⓒ Ⓓ   36. Ⓐ Ⓑ Ⓒ Ⓓ
3. Ⓐ Ⓑ Ⓒ Ⓓ   20. Ⓐ Ⓑ Ⓒ Ⓓ   37. Ⓐ Ⓑ Ⓒ Ⓓ
4. Ⓐ Ⓑ Ⓒ Ⓓ   21. Ⓐ Ⓑ Ⓒ Ⓓ   38. Ⓐ Ⓑ Ⓒ Ⓓ
5. Ⓐ Ⓑ Ⓒ Ⓓ   22. Ⓐ Ⓑ Ⓒ Ⓓ   39. Ⓐ Ⓑ Ⓒ Ⓓ
6. Ⓐ Ⓑ Ⓒ Ⓓ   23. Ⓐ Ⓑ Ⓒ Ⓓ   40. Ⓐ Ⓑ Ⓒ Ⓓ
7. Ⓐ Ⓑ Ⓒ Ⓓ   24. Ⓐ Ⓑ Ⓒ Ⓓ   41. Ⓐ Ⓑ Ⓒ Ⓓ
8. Ⓐ Ⓑ Ⓒ Ⓓ   25. Ⓐ Ⓑ Ⓒ Ⓓ   42. Ⓐ Ⓑ Ⓒ Ⓓ
9. Ⓐ Ⓑ Ⓒ Ⓓ   26. Ⓐ Ⓑ Ⓒ Ⓓ   43. Ⓐ Ⓑ Ⓒ Ⓓ
10. Ⓐ Ⓑ Ⓒ Ⓓ   27. Ⓐ Ⓑ Ⓒ Ⓓ   44. Ⓐ Ⓑ Ⓒ Ⓓ
11. Ⓐ Ⓑ Ⓒ Ⓓ   28. Ⓐ Ⓑ Ⓒ Ⓓ   45. Ⓐ Ⓑ Ⓒ Ⓓ
12. Ⓐ Ⓑ Ⓒ Ⓓ   29. Ⓐ Ⓑ Ⓒ Ⓓ   46. Ⓐ Ⓑ Ⓒ Ⓓ
13. Ⓐ Ⓑ Ⓒ Ⓓ   30. Ⓐ Ⓑ Ⓒ Ⓓ   47. Ⓐ Ⓑ Ⓒ Ⓓ
14. Ⓐ Ⓑ Ⓒ Ⓓ   31. Ⓐ Ⓑ Ⓒ Ⓓ   48. Ⓐ Ⓑ Ⓒ Ⓓ
15. Ⓐ Ⓑ Ⓒ Ⓓ   32. Ⓐ Ⓑ Ⓒ Ⓓ   49. Ⓐ Ⓑ Ⓒ Ⓓ
16. Ⓐ Ⓑ Ⓒ Ⓓ   33. Ⓐ Ⓑ Ⓒ Ⓓ   50. Ⓐ Ⓑ Ⓒ Ⓓ
17. Ⓐ Ⓑ Ⓒ Ⓓ   34. Ⓐ Ⓑ Ⓒ Ⓓ

Test 1 Answer Key

1. A	**11.** D	**21.** C	**31.** D	**41.** A
2. B	**12.** D	**22.** D	**32.** D	**42.** C
3. C	**13.** A	**23.** B	**33.** B	**43.** A
4. C	**14.** C	**24.** A	**34.** A	**44.** C
5. A	**15.** A	**25.** D	**35.** C	**45.** D
6. D	**16.** B	**26.** C	**36.** C	**46.** A
7. B	**17.** C	**27.** C	**37.** B	**47.** B
8. B	**18.** B	**28.** B	**38.** A	**48.** D
9. D	**19.** A	**29.** C	**39.** C	**49.** B
10. C	**20.** B	**30.** A	**40.** D	**50.** D

Answer Key Explanations

1. Correct response: **A**
(*Compare and order whole numbers*)
   To compare and order whole numbers, use place value to find the value of the digits. Since these numbers all have seven digits, you can find the greatest number by looking at the first digit in each number: **9**,072,576 is the greatest number.

Incorrect choices:

B, **C**, and **D** are all less than 9,072,576.

2. Correct response: **B**
(*Compare, order, and use percents*)
   To compare the sale prices, apply the discount to the original price of the helmet in each store. On sale, the helmets are least expensive at Bikeway: 50% of $60 is $30 ($0.5 \times \$60 = \$30$); the sale price is $60 − $30 = $30.

2. (continued)
  Incorrect choices:

A 10% of $40 is $4; the sale price is $40 − $4 = $36.

C 20% of $50 is $10; the sale price is $50 − $10 = $40.

D 10% of $45 is $4.50; the sale price is $45 − $4.50 = $40.50.

3. Correct response: **C**
(*Identify and use place value*)
   This number in expanded form is 2,000,000 + 700,000 + 60,000 + **4**,000 + 70 + 5. The **4** represents the thousands place.

Incorrect choices:

A The 0 represents the hundreds place in this number.

B The 2 represents the millions place.

D The 6 represents the ten thousands place.

4. Correct response: **C**
(*Compare, order, and use integers*)

To compare and order integers, you can make a number line or determine which number is closer to 0. Negative integers are less than 0 and are located to the left of the 0 on a number line. The integer –69 is farthest from 0 and has the least value; it represents the lowest temperature.

Incorrect choices:

A -55 is less than -36 but greater than -66 and -69.

B -66 is less than -36 and -55 but greater than -69.

D -36 is closest to 0 and greater than -55, -69, and -66.

5. Correct response: **A**
(*Identify prime and composite numbers and factors*)

The numbers in this group are all prime numbers because they can only be factored, or divided evenly, by 1 and themselves.

Incorrect choices:

B includes two prime numbers (5 and 29), but 33 and 45 are composite numbers (3 and 11 are factors of 33; 3, 5, 9, and 15 are factors of 45).

C includes two prime numbers (3 and 17), but 8 and 12 are composite numbers (2 and 4 are factors of 8; 2, 3, 4, and 6 are factors of 12).

D includes two prime numbers (11 and 19), but 16 and 35 are composite numbers (2, 4, and 8 are factors of 16; 5 and 7 are factors of 35).

6. Correct response: **D**
(*Use expanded notation*)

The expanded form of a number shows the place value of each digit, so the correct form of 14,858 is 10,000 + 4,000 + 800 + 50 + 8.

6. (continued)
Incorrect choices:

A incorrectly shows 14,000 (instead of 10,000 + 4,000) and shows 858 as 850 + 8.

B incorrectly shows 14,000 (instead of 10,000 + 4,000).

C incorrectly shows 4,800 (instead of 4,000 + 800).

7. Correct response: **B**
(*Use ratios to describe and compare two sets of data*)

To find the ratio, compare the part (number of free throws made) to the whole (total number of free throws): 8 to 12, or $\frac{8}{12}$; then reduce $\frac{8}{12}$ to $\frac{2}{3}$.

Incorrect choices:

A The ratio $\frac{12}{8}$ represents the whole to the part and not the part to the whole.

C The ratio $\frac{3}{4}$ represents an error in reducing $\frac{8}{12}$.

D The ratio $\frac{8}{20}$ mistakenly assumes that Marcus made 8 free throws and missed 12, so the total number of attempts was 20.

8. Correct response: **B**
(*Add, subtract, multiply, and divide whole numbers*)

This is a two-step problem. To find the total number of players, multiply 4 teams by 18 players per team: $4 \times 18 = 72$. Then subtract the 20 players who are 12 years old from the total: $72 - 20 = 52$ players who are younger than 12.

Incorrect choices:

A is the number of players who are 12 years old.

C reflects an error in multiplying and then subtracting.

D reflects an estimated total of 80 players minus the number of 12-year-olds.

9. Correct response: **D**
(*Add, subtract, multiply, and divide mixed numbers*)

To determine the difference in speed, subtract $17 \frac{4}{5}$ from $33 \frac{3}{10}$. Using a common denominator of 10, you can rename and subtract: $32 \frac{13}{10} - 17 \frac{8}{10} = 15 \frac{5}{10}$, or $15 \frac{1}{2}$.

9. (continued)

Incorrect choices:

A is the result of adding $33\frac{3}{10} + 17\frac{4}{5}$ instead of subtracting.

B reflects an error in renaming the mixed numbers before subtracting.

C reflects an error in subtracting fractions.

10. Correct response: **C**
(*Add, subtract, multiply, and divide decimals*)
To find the total amount of cheese produced, add the amount produced each week: $46.7 + 41.4 + 40.2 + 40.1 = 168.4$ lbs.

Incorrect choices:

A is the result of adding the four amounts (168.4) and dividing by 4 to find the average amount per week.

B is the result of not regrouping (or carrying the 1) after adding the first column of numbers.

D is the result of multiplying the amount in the first week (46.7 lbs) by 4 weeks.

11. Correct response: **D**
(*Solve problems involving integers*)
To determine who scored 104 after two rounds, add or subtract each player's scores from the par of 54. Then add the scores from the two rounds. Linda scored 104: $54 - 4 = 50$ in the first round and $54 + 0 = 54$ in the second round: $50 + 54 = 104$.

Incorrect choices:

A Chris scored $54 + 0$ in the first round and $54 + 0$ in the second round, for a total of 108.

B Melinda scored $54 + 3 = 57$ in the first round and $54 - 2 = 52$ in the second round, for a total of 109.

C Joanna scored $54 - 5 = 49$ in the first round and $54 - 1 = 53$ in the second round, for a total of 102.

12. Correct response: **D**
(*Estimate and round using whole numbers*)
To estimate the total number of seats, add 28 tables + 9 more tables ($28 + 9 = 37$), and round the sum to 40. Round the number of seats per table from 11 to 10, and then multiply: $40 \times 10 = 400$.

Incorrect choices:

A is the result of failing to add 9 more tables and rounding 28 down to 25: $25 \times 10 = 250$.

B is the result of failing to add 9 more tables and rounding up to 30: $30 \times 10 = 300$.

C is the result of not rounding 37 upward: $37 \times 10 = 370$.

13. Correct response: **A**
(*Use order of operations to simplify whole number expressions*)
To simplify, complete the operations in parentheses first: $(15 \div 3) = 5$, and $(4 \times 5) = 20$. Then subtract and add from left to right: $5 - 2 + 20 = 23$.

Incorrect choices:

B is the result of adding $5 + 20$, forgetting to subtract 2.

C is the result of adding the 2 instead of subtracting: $5 + 2 + 20 = 27$.

D is the result of multiplying 5×2 instead of subtracting $5 - 2$: $10 + 20 = 30$.

14. Correct response: **C**
(*Apply the properties of operations*)
The inverse operation of division is multiplication; since $320 \div 40 = 8$, then the inverse is 40×8.

Incorrect choices:

A $320 + 40$ is the inverse operation of subtraction, not division.

B Division is not commutative, so $40 \div 320$ does not equal $320 \div 40$.

D $320 - 40$ is the inverse operation of addition, not division.

15. Correct response: **A**

(*Solve multi-step problems involving whole numbers and decimals*)

To find how much Dora spent on flowers, subtract the amount she spent on paper products and balloons ($15.50 + $12) from the total of $50 spent: $50 − $27.50 = $22.50.

Incorrect choices:

B The number sentence adds the amounts spent on paper products and balloons but does not subtract that sum from $50.

C The number sentence subtracts $12 from $15.50 instead of adding these two amounts.

D The number sentence adds $15.50 and $12 to the total of $50 instead of subtracting.

16. Correct response: **B**

(*Solve multi-step problems involving whole numbers*)

To find the number of cards Jake collected in the third month, subtract the number of cards he collected in the first two months (45 + 27 = 72) from the total number of cards (110): 110 − 72 = 38.

Incorrect choices:

A reflects an error in regrouping when subtracting 110 − 72.

C is the result of subtracting 110 − 45, the number of cards collected in the first month.

D is the result of subtracting 110 − 27, the number of cards collected in the second month.

17. Correct response: **C**

(*Solve number sentences with one variable*)

To find the value of this expression, substitute the number (7) for the variable (c) and then compute: $\frac{4 + (8 \times 7)}{3}$. Since $8 \times 7 = 56$, then $4 + 56 = 60$; $60 \div 3 = 20$.

17. (continued)

Incorrect choices:

A is the result of adding 4 + 8 + 7 (instead of multiplying 8 × 7) and dividing by 3.

B is the result of multiplying 8 × 7 = 56 and dividing by 3, forgetting to add the 4 to 56.

D is the result of adding 4 + 56 = 60 but not dividing by 3.

18. Correct response: **B**

(*Convert or estimate conversions of measures*)

Since 1 foot = 12 inches, then you can convert 5 feet to inches by multiplying: 5 × 12 = 60 inches. Sam is 5 feet 2 inches, or 62 inches, tall. Andy is 4 inches shorter. To find Andy's height, subtract: 62 − 4 = 58 inches.

Incorrect choices:

A is the result of subtracting 60 − 4, forgetting to add the 2 inches to Sam's height.

C is the result of converting 5 feet to inches, but forgetting to add 2 inches and then subtract 4 inches.

D is the result of converting 5 feet to inches and then adding 4 inches instead of subtracting.

19. Correct response: **A**

(*Select appropriate unit for measuring length*)

Most house keys are about 2 inches long. An inch is about 2.5 centimeters, so 5 centimeters is a reasonable measurement for the length of a house key.

Incorrect choices:

B Kilograms are used to measure weight/mass, not length.

C A meter is about one yard, or the width of a door; 5 meters would be too long for a key.

D A millimeter is smaller than a centimeter; 5 millimeters would be too short for a key.

20. Correct response: **B**

(*Estimate and find length and perimeter*)

The perimeter is the distance around a figure. Since a rectangle has two pairs of equal sides, you can find the perimeter by adding two lengths and two widths: 8 ft + 8 ft + 6 ft + 6 ft = 28 ft.

20. (continued)

Incorrect choices:

A is the result of adding only one length and one width: 8 ft + 6 ft.

C reflects an error in adding or multiplying the lengths and widths (16 ft + 14 ft).

D is the result of multiplying to find the area of the garden (8 ft × 6 ft) instead of the perimeter.

21. Correct response: **C**

(*Estimate and find length and perimeter*)

The perimeter is the distance around the outside of the room. The room is rectangular, so it has two pairs of equal sides. The width of the room is 8 feet, so two of the walls measure 8 feet: 8 ft × 2 = 16 ft. Subtract 16 from the perimeter (36) to find the length of the other two walls: 36 − 16 = 20 ft. Divide by 2 to find the length of one wall: 20 ÷ 2 = 10 ft.

Incorrect choices:

A is the combined length of two walls (36 − 16), which should be divided by 2.

B is the combined width of two walls: 8 + 8.

D is the width of the room, which would only be the same as the length if the room were a square.

22. Correct response: **D**

(*Identify and describe the radius and diameter of a circle*)

The diameter is the width of a circle; it is defined by a chord that passes through the center of the circle. Line segment *AB* is the only segment that fits the description of a diameter.

Incorrect choices:

A Line segment OC is a radius, which is $\frac{1}{2}$ the length of a diameter.

B Line segment OD is a radius, which is $\frac{1}{2}$ the length of a diameter.

C Line segment DE is a chord, a line segment that has its endpoints on the circle.

23. Correct response: **B**

(*Classify acute, obtuse, and right angles and triangles*)

An angle greater than 90° is called obtuse; a triangle with an angle greater than 90° is an obtuse triangle.

Incorrect choices:

A An acute triangle has three angles that measure less than 90°.

C A right triangle has one right angle, which measures exactly 90°.

D An equilateral triangle has three equal sides.

24. Correct response: **A**

(*Identify, classify, and describe plane figures and their attributes*)

The figure shown is an octagon because it has eight sides (the prefix *octa-* means "eight").

Incorrect choices:

B A rhombus is a parallelogram with four congruent sides.

C A hexagon has six sides.

D A pentagon has five sides.

25. Correct response: **D**

(*Determine congruence and similarity*)

Similar figures have the same shape but may not be the same size. These two figures are similar isosceles triangles (with two equal sides) but are different sizes.

Incorrect choices:

A The figure is a right triangle; it is not the same shape or size.

B The figure is a scalene triangle; it is not the same shape and has no equal sides.

C The figure is an equilateral triangle; it is not the same shape and has three equal sides.

26. Correct response: **C**
(*Identify, classify, and describe solid figures and their attributes*)

This figure is a cylinder because it has two flat, parallel, congruent circular bases and a curved lateral surface.

Incorrect choices:

A A cone is a solid, pointed figure with one flat, round base.

B A pyramid is a solid figure with a polygon base and triangular sides.

D A prism is a solid figure formed by polygons.

27. Correct response: **C**
(*Locate and name points on the coordinate plane using ordered pairs*)

In an ordered pair, the first number is the *x*-coordinate, going across to the left or the right. The second number is the *y*-coordinate, going up or down. On the coordinate plane, Mount Snow is located at 4 across and 6 up, or (4, 6).

Incorrect choices:

A (1, 4) is the location of Magic Mt.

B (4, 4) is the location of Colony.

D confuses the *x*-coordinate and the *y*-coordinate; (6, 4) is the location of Ray's Peak.

28. Correct response: **B**
(*Locate and name points on coordinate plane using ordered pairs*)

In an ordered pair, the first number is the *x*-coordinate, going across to the left or the right. The second number is the *y*-coordinate, going up or down. The point described by the ordered pair (3, 5) is 3 across and 5 up; that is the location of Pat's Hut.

Incorrect choices:

A reflects a confusion between the *x*-coordinate and the *y*-coordinate; Walli's Warming Hut is located at (5, 3).

C The location of Damon's Hut is 6 across and 4 up, or (6, 4).

D The location of Al's Hideaway is 2 across and 3 up, or (2, 3).

29. Correct response: **C**
(*Identify and describe points, lines, line segments, and rays*)

A line segment is part of a line with two endpoints and all the points between them.

Incorrect choices:

A is a line, not a line segment, because it has no endpoints; it is a straight path of points that continues without end in both directions.

B is a point, not a line segment, because it names an exact location in space.

D is a ray, not a line segment, because it has one endpoint; it continues without end in one direction, and it is part of a line.

30. Correct response: **A**
(*Identify lines of symmetry*)

A figure has a line of symmetry when it can be folded or divided into two parts that are congruent. The letter T has one line of symmetry; it can be divided by a vertical line drawn straight down the middle.

Incorrect choices:

B The letter S has no lines of symmetry because it cannot be divided into two equal parts.

C and **D** reflect a misunderstanding of symmetry.

31. Correct response: **D**
(*Transform figures in the coordinate plane*)

To determine transformations, look at the movement of the figures to tell whether they slide, flip, or turn. A rotation is a turn; the second figure in this pair has turned 90° around a point.

Incorrect choices:

A is a reflection; the second figure of the pair has flipped over a line and now presents a mirror image of the first figure.

B is a translation; the second figure of the pair has moved across to a new position along a straight line.

C is a translation and a dilation; the second figure of the pair has moved to the right and has been enlarged.

32. Correct response: **D**
(*Solve problems involving length, perimeter, and area*)

The patio is a square, so it has four sides of equal length. To find the length of each side, divide: $32 \div 4 = 8$ ft. To find the area of a rectangle, multiply the length times the width. In a square, the length and width are the same: 8 ft $\times 8$ ft $= 64$ square feet.

Incorrect choices:

A represents the length of one side: 32 ft $\div 4 = 8$ ft.

B is the result of adding two sides (8 ft $+ 8$ ft) instead of multiplying.

C represents the perimeter of the square, or the sum of all four sides.

33. Correct response: **B**
(*Solve problems involving time*)

To find what time Sara leaves, add the amounts of time to, or count forward from, her starting time: $9{:}30$ A.M. $+ 40$ minutes $= 10{:}10$ A.M.; $10{:}10$ A.M. $+ 45$ minutes $= 10{:}55$ A.M.; $10{:}55$ A.M. $+ 1$ hour 20 minutes $= 12{:}15$ P.M.

Incorrect choices:

A reflects an error of 10 minutes in calculating the time, adding 30 minutes for the first task instead of 40 minutes.

C reflects an error of 5 minutes in calculating the time, adding 1 hour 20 minutes to $11{:}00$ A.M. instead of $10{:}55$.

D reflects an error of 10 minutes in calculating the time, adding 50 minutes for the first task instead of 40 minutes.

34. Correct response: **A**
(*Solve problems involving proportions*)

To find the height of the model, you can set up a proportion: $\dfrac{1 \text{ in.}}{250 \text{ ft.}} = \dfrac{h}{1{,}250 \text{ ft.}}$. Then solve the proportion by multiplying cross products: $250h = 1{,}250$ ft. To isolate the h on one side of the equation, divide each side by 250: $h = 1{,}250 \div 250$, or $h = 5$. The model should be 5 inches tall.

34. (continued)
Incorrect choices:

B reflects an error in dividing $1{,}250$ by 250.

C and **D** reflect an incorrect proportion or a misunderstanding of how to use proportions.

35. Correct response: **C**
(*Interpret data presented in a bar graph*)

The bar graph shows that a wild turkey can run 15 miles per hour. An animal that runs 3 times that fast (3×15) would have to run 45 miles per hour. The graph shows that an elk runs 45 miles per hour.

Incorrect choices:

A is the result of misreading the graph; the crocodile runs 5 miles per hour slower than the wild turkey.

B reflects a misinterpretation of the graph; the elephant runs 25 miles per hour, or 10 miles per hour faster than the wild turkey.

D is the result of misreading the graph; the ostrich runs 40 miles per hour, but that is not 3 times as fast as the wild turkey.

36. Correct response: **C**
(*Interpret data presented in a circle graph*)

The percentage of students who *did* choose orange or grapefruit juice was $40\% + 4\% = 44\%$. To find the percentage of students who did *not* choose either of these two juices, subtract this amount from 100%: $100\% - 44\% = 56\%$.

Incorrect choices:

A reflects the number of students who chose orange or grapefruit juice ($40\% + 4\%$).

B reflects the number of students who chose orange or pineapple juice ($40\% + 6\%$).

D reflects the number of students who chose any juice other than orange juice ($100\% - 40\%$).

37. Correct response: **B**

(*Interpret data in stem-and-leaf plots*)

The easiest way to find the correct stem-and-leaf plot is to list the ages in the box by tens in sequence:

43, 46
51, 52, 54, 55, 56
60, 61, 62, 64, 69

Then you can compare these numbers to the representation in the stem-and-leaf plot.

Incorrect choices:

A The stems (4, 5, 6) are correct, but the leaves do not match the data.

C The stems are correct, but the leaves for 4 and 6 are reversed.

D The leaves list the number of items that go with each stem instead of listing the 12 data points.

38. Correct response: **A**

(*List all possible outcomes or construct sample spaces using lists, charts, and tree diagrams*)

There are three sandwiches, two fruits, and two drinks, so there are 12 possible combinations of one sandwich, one fruit, and one drink: $3 \times 2 \times 2 = 12$. Choice A is the only tree diagram that lists 12 combinations correctly.

Incorrect choices:

B is incomplete. It lists the choices from the chart in the same order as they are presented.

C is incomplete. It lists the correct combinations of sandwiches and fruit, but it does not list two drink possibilities with each combination.

D lists all of the possible choices and connects them all with one another, but this kind of diagram does not show all the possible combinations.

39. Correct response: **C**

(*Find probabilities, represented as ratios*)

There are eight letters in all, so there will be eight index cards in the bag. Since four of the letters are vowels (*A, A, O, O*), then 4 of the 8 possible choices are vowels. The probability of choosing a vowel in one try is 4 out of 8, or $\frac{4}{8}$, or $\frac{1}{2}$.

39. (continued)

Incorrect choices:

A is the probability of choosing any one of the cards out of eight cards.

B is the probability of choosing an *A* or an *O*, if you assume that there is only one *A* card and one *O* card.

D reflects an error in counting the number of vowels (a total of five instead of four).

40. Correct response: **D**

(*Determine and compare probabilities for simple and compound events*)

There are 20 marbles in the bag all together: $9 + 6 + 4 + 1 = 20$. The probability of choosing a red marble is $\frac{9}{20}$; a blue marble is $\frac{6}{20}$; a green marble is $\frac{4}{20}$, and a yellow marble is $\frac{1}{20}$. Since there is only one yellow marble in the bag out of 20 marbles, Ken is least likely to get a yellow marble.

Incorrect choices:

A is the color Ken is most likely to get; the probability of choosing a red marble is $\frac{9}{20}$.

B and **C** are less likely choices, but both are still more likely than yellow.

41. Correct response: **A**

(*Determine and describe the mean, median, mode, and range of data*)

To find the range of temperatures in this table, subtract the lowest temperature (104°) from the highest temperature (134°): $134° - 104° = 30°$.

Incorrect choices:

B represents the mean temperature.

C represents the median temperature.

D is the highest temperature listed in the table.

42. Correct response: **C**

(*Determine and describe the mean, median, mode, and range of data*)

To find the median, list the numbers in numerical order: 104, 109, 115, 120, 125, 128, 134. The median is the number in the middle of the set of numbers: 120.

42. (continued)
Incorrect choices:

A represents the lowest number in the set.

B reflects an error in finding the middle number; 115 is the third number of seven.

D represents the greatest number in the set.

43. Correct response: **A**
(*Collect, organize, display, and interpret data to solve problems*)

A bar graph is most appropriate for displaying and comparing discrete data. In this case, each bar on the graph would represent a number of representatives from one state.

Incorrect choices:

B A circle graph (or pie chart) is not a good choice because it shows data as parts of a whole.

C A stem-and-leaf plot is not a good choice because it is best for showing data related to only one factor (such as the heights of students).

D A line graph is not a good choice because it shows data changing over a period of time.

44. Correct response: **C**
(*Identify, describe, and extend numerical and geometric patterns*)

To determine what the 11th shape will be, you must figure out the pattern and then extend it to eleven shapes. The pattern shows a white square–triangle–trapezoid, then a black square and triangle. Based on this pattern, the next shape will be a black trapezoid. Then the six shapes (three white and three black) will repeat, and the 11th shape will be a black triangle.

Incorrect choices:

A is the correct shape, but the triangle is white; it should be black.

B reflects an error in counting the shapes; this is the 10th shape in the pattern.

D reflects an error in counting the shapes; this is the 9th shape in the pattern.

45. Correct response: **D**
(*Interpret, write, and simplify algebraic expressions*)

The question says that Sharon collected b number of bottles, and Kayla collected "32 more than twice the number of bottles Sharon collected." Twice the number of bottles Sharon collected can be expressed as $2 \times b$, or $2b$. 32 more than that number is $32 + 2b$.

Incorrect choices:

A The expression uses the wrong operation, subtracting instead of adding; $32 - 2b$ represents "32 minus twice the number of bottles."

B The expression adds 2 bottles instead of multiplying the number of bottles by 2; $2 + (32 + b)$ represents "2 plus 32 plus the number of bottles."

C The expression multiplies 32 times $2b$ instead of adding 32.

46. Correct response: **A**
(*Apply basic properties and order of operations with algebraic expressions*)

Mario buys 3 cornflower plants for $9.95 each. The cost can be expressed as 3×9.95. He bought 3 black-eyed Susans at m each. This cost can be represented as $3 \times m$. Using the distributive property, these two expressions can be combined as $3 \times (9.95 + m)$ to represent the total amount he spent.

Incorrect choices:

B The expression uses the wrong operation, adding instead of multiplying.

C The expression adds the costs of two plants and divides by 3 instead of multiplying.

D The expression multiplies the cost of 3 cornflowers times the cost of 3 black-eyed Susans instead of adding the two costs.

47. Correct response: **B**

(*Use simple equations to represent problem situations*)

The number sentence has a variable (n), meaning that one of the numbers is unknown. Beth made 54 bracelets and gave some away. She now has 41 bracelets left. Using n to represent the number of bracelets she gave away, you can describe this situation as $54 - n = 41$.

Incorrect choices:

A There is no unknown quantity; the situation can be represented as $41 + 13 = 54$.

C This situation can be represented by the expression $54 + 13 + 41$; it does not involve an equation.

D This situation does have an unknown quantity (the number of friends), but it requires division (dividing 54 sandwiches); it does not fit the number sentence $54 - n = 41$.

48. Correct response: **D**

(*Represent and describe mathematical relationships with tables*)

To find the cost of the bike rental for 8 hours, find the number pattern in the table. Since the cost increases by $3 for each hour, the pattern is "+ 3." The cost for 4 hours is $17, so the cost for 8 hours is $17 + \$3 + \$3 + \$3 + \$3 = \$29$. Then add the cost of the bike helmet: $\$29 + \$8 = \$37$.

Incorrect choices:

A Represents the cost of the bike rental for 4 hours ($17) plus the cost of the bike helmet ($8).

B Represents the cost of the bike rental for 8 hours but does not include the cost of the bike helmet.

C Represents the cost of the bike rental for 8 hours based on a pattern of "+ 2" instead of "+ 3": $\$25 + \$8 = \$33$.

49. Correct response: **B**

(*Solve one-step linear equations*)

To solve for x, you must first isolate x on one side of the equation. To do this, subtract 7 from both sides of the equation, and then divide both sides by 5:

$$5x + 7 (- 7) = 42 (- 7), \text{ or } 5x = 35$$

$$\frac{5x}{5} = \frac{35}{5}, \text{ or } x = 7.$$

Incorrect choices:

A reflects an error in division: $\frac{35}{5}$ does not equal 8.

C reflects an error in division: $\frac{35}{5}$ does not equal 6.

D reflects an error in addition: if $x = 5$, then $5x = 25$; but $25 + 7 = 32$, not 42.

50. Correct response: **D**

(*Make predictions from patterns, data, or examples*)

To find the number of push-ups Janice will do daily in Week 9, you must first find the pattern in the number of push-ups and then extend the table to Week 9. Since Janice adds 4 push-ups each week ($6 + 4 = 10$), then the pattern is "+ 4." Since she does 18 push-ups each day in Week 4, the number of push-ups in Week 9 will be $18 + 4 + 4 + 4 + 4 + 4 = 38$.

Incorrect choices:

A is the number of push-ups she will do daily in Week 5 ($18 + 4 = 22$).

B is the number of push-ups she will do daily in Week 7 ($18 + 4 + 4 + 4 = 30$).

C is the number of push-ups she will do daily in Week 8 ($18 + 4 + 4 + 4 + 4 = 34$).

Test ② Answer Key

1. B	**11.** C	**21.** C	**31.** C	**41.** B
2. C	**12.** B	**22.** C	**32.** D	**42.** C
3. A	**13.** D	**23.** B	**33.** B	**43.** B
4. D	**14.** A	**24.** D	**34.** C	**44.** D
5. B	**15.** B	**25.** C	**35.** D	**45.** A
6. D	**16.** C	**26.** A	**36.** D	**46.** C
7. A	**17.** C	**27.** B	**37.** C	**47.** D
8. D	**18.** A	**28.** D	**38.** B	**48.** B
9. B	**19.** D	**29.** C	**39.** B	**49.** B
10. C	**20.** D	**30.** A	**40.** A	**50.** D

Answer Key Explanations

1. Correct response: **B**
(*Compare and order whole numbers and fractions*)
To compare and order fractions and whole numbers, find the common denominator and convert the fractions. Using eighths as the common denominator, the fractions can be converted: $\frac{1}{2} = \frac{4}{8}$ and $\frac{1}{4} = \frac{2}{8}$. Then sequence the amounts from least to greatest: $\frac{1}{8}, \frac{2}{8}, \frac{4}{8}, \frac{8}{8}$, or 1. The $\frac{1}{8}$ teaspoon of red pepper is the least amount.

Incorrect choices:

A $\frac{4}{8}$ teaspoon of oregano is greater than $\frac{1}{8}$.

C $\frac{2}{8}$ teaspoon of basil is greater than $\frac{1}{8}$.

D $1 = \frac{8}{8}$; it represents the greatest amount.

2. Correct response: **C**
(*Compare, order, and use percents*)
To determine how many students attended school each week, apply the percentage to the total number of students at the school (600). For Week 3, 80% of the students went to school: $0.8 \times 600 = 480$, and 480 students is less than 500.

2. (continued)
Incorrect choices:

A 84% of 600 (0.84×600) = 504, which is greater than 500.

B 88% of 600 (0.88×600) = 528, which is greater than 500.

D 92% of 600 (0.92×600) = 552, which is greater than 500.

3. Correct response: **A**
(*Identify and use place value*)
This number represents $3,000,000 + \mathbf{900,000} + 80,000 + 1,000 + 500 + 40 + 6$. The **9** represents the hundred thousands place.

Incorrect choices:

B The 8 represents the ten thousands place.

C The 5 represents the hundreds place.

D The 3 represents the millions place.

4. Correct response: **D**

(*Compare, order, and use integers*)

In relation to elevation, sea level represents 0 feet. To compare and order these integers, decide which one is closest to 0. For negative integers, the larger the absolute value of the number, the lesser the value. An elevation of −512 is 512 feet below sea level. The elevation closest to sea level is −52.

Incorrect choices:

A −512 is the lowest elevation listed and is farthest from 0.

B −132 is 132 feet below sea level, and that is lower than −52.

C −92 is 92 feet below sea level, and that is lower than −52.

5. Correct response: **B**

(*Identify prime and composite numbers*)

A composite number is a whole number that has more than two whole-number factors. Of the numbers listed, only 15 has more than two factors (1, 3, 5, 15).

Incorrect choices:

A, **C**, and **D** are all prime numbers; each of them has only two factors (for example, the factors of 13 are 1 and 13).

6. Correct response: **D**

(*Use expanded notation*)

The expanded form of a number shows the place value of each digit, so the correct form of 375,291 is 300,000 + 70,000 + 5,000 + 200 + 90 + 1.

Incorrect choices:

A incorrectly shows 370,000 (instead of 300,000 + 70,000) and shows 291 as 290 + 1.

B incorrectly shows 370,000 (instead of 300,000 + 70,000) and shows 5,000 as 50,000.

C incorrectly shows 5,000 as 50,000.

7. Correct response: **A**

(*Use ratios to describe and compare two sets of data*)

To find the ratio, compare the part (number of games won = 10) to the whole (total number of games played = 10 + 6); the ratio is 10 to 16, or 10:16.

Incorrect choices:

B The ratio 4:16 represents an error in the number of games won (10 games − 6 games lost = 4 games).

C The ratio 10:10 represents the number of games won to the number of games won, not to the total number of games.

D The ratio 10:6 represents the number of games won to the number of games lost, not to the total number of games (10 + 6).

8. Correct response: **D**

(*Add, subtract, multiply, and divide whole numbers*)

To find Kyle's age (k), you can write a number sentence. You know that Adrian is 14 years old. Manny is 3 years older than Kyle, so you can represent Manny's age as $k + 3$. The sum of their ages is 41, so: $14 + (k + 3) + k = 41$.

Solve this equation by simplifying: $17 + 2k = 41$.

Then isolate k by subtracting 17 from each side of the equation:

$17 (- 17) + 2k = 41 (- 17)$, or $2k = 24$.

Then divide each side by 2: $\dfrac{2k}{2} = \dfrac{24}{2}$, or $k = 12$.

Incorrect choices:

A is the result of not dividing by 2 in the final step ($2k = 24$).

B is the result of adding 3 years to Adrian's age ($14 + 3$).

C is Manny's age rather than Kyle's ($12 + 3 = 15$).

9. Correct response: **B**

(*Add, subtract, multiply, and divide fractions and mixed numbers*)

To find the total number of votes needed, you can add the number of people in each group ($100 + 435 = 535$) and multiply by $\frac{2}{3}$: $535 \times \frac{2}{3} = \frac{1,070}{3} = 356\frac{2}{3}$. The total number of votes needed is about 356.

Incorrect choices:

A represents the total number of people in each group: $100 + 435$.

C represents $\frac{2}{3}$ of the number of Representatives: $435 \times \frac{2}{3}$.

D represents about $\frac{1}{3}$ of the total number of people: $\frac{1}{3} \times 535$.

10. Correct response: **C**

(*Add, subtract, multiply, and divide decimals*)

To find the total weight of these three whales, add the three weights: $49.6 + 41.1 + 39.7 = 130.4$ tons.

Incorrect choices:

A is the result of adding only two of the weights ($41.1 + 39.7$).

B is the result of adding only the first two weights ($49.6 + 41.1$).

D is the result of an error in addition.

11. Correct response: **C**

(*Solve problems involving integers*)

According to the table, the temperature in Point Barrow is $-12.1°$. To find which temperature is $7.6°$ colder (or lower) than Point Barrow, add $-7.6°$ to $-12.1°$. The result is $-19.7°$, the temperature in Eureka, Canada.

Incorrect choices:

A is the result of adding $7.6°$ and $-12.1°$ (instead of $-7.6°$ and $-12.1°$).

B is the temperature in Ostrov Bolshoy.

D is the temperature in Resolute, Canada.

12. Correct response: **B**

(*Estimate and round using whole numbers and decimals*)

To estimate the total length of all four tunnels, round the length of each tunnel to the nearest tens place and then add the numbers: $20 + 10 + 50 + 20 = 100$ km.

Incorrect choices:

A is the result of rounding all the numbers to the nearest 5: $20 + 5 + 50 + 15 = 90$.

C is the result of rounding 22.2 to 30 and 53.8 to 60, and rounding all the numbers upward: $30 + 10 + 60 + 20 = 120$.

D is too high.

13. Correct response: **D**

(*Use order of operations to simplify whole-number expressions*)

To simplify, complete the operations in parentheses first: $(36 \div 3) = 12$, and $(5 \times 7) = 35$. Then add and subtract from left to right: $12 + 25 = 37$, and $37 - 35 = 2$.

Incorrect choices:

A is the result of calculating from left to right, disregarding the parentheses.

B is the result of adding $37 + 35$ instead of subtracting.

C is the result of subtracting $37 - 5$ and omitting the last number ($\times 7$).

14. Correct response: **A**

(*Apply the properties of operations*)

The inverse operation of subtraction is addition; since $45 - 36 = 9$, then the inverse is $36 + 9 = 45$.

Incorrect choices:

B Multiplication is not the inverse of subtraction.

C Adding the two numbers together is not the inverse of subtracting them.

D 45×36 is the inverse operation of division, not subtraction.

15. Correct response: **B**

(*Solve multi-step problems involving whole numbers*)

　　To find how many students voted for other candidates, subtract the number of students who voted for Mark from the total number of students. Mark received $\frac{3}{5}$ of the 30 votes: $\frac{3}{5} \times 30 = \frac{90}{5} = 18$ votes for Mark. $30 - 18 = 12$ students who voted for other candidates.

Incorrect choices:

A is the result of an error in computation, resulting in $\frac{1}{3}$ of 30.

C is the result of trying to find $\frac{2}{5}$ of 30 by using the numerator to divide: $30 \div 2 = 15$.

D is the number of students who voted for Mark ($\frac{3}{5} \times 30$).

16. Correct response: **C**

(*Solve multi-step problems involving whole numbers and decimals*)

　　To find how much Carly earned in 4 weeks, multiply the number of hours she worked each week (12) times her hourly wage ($8.25) $\times$ the number of weeks (4): $12 \times \$8.25 = \$99$; $\$99 \times 4 = \$396$.

Incorrect choices:

A is the amount she earned in only one week: 12 hours $\times$ $8.25.

B is the result of doubling her earnings for one week instead of multiplying by 4 ($2 \times \$99 = \$198$).

D is the result of rounding $8.25 to $9 ($12 \times \$9 = \$108 \times 4$ weeks $= \$432$).

17. Correct response: **C**

(*Solve number sentences with one variable*)

　　To find the value of this expression, substitute the number 7 for the variable (b) and then compute: $(4 \times 7) + (6 + 7) = 28 + 13 = 41$.

17. (continued)

Incorrect choices:

A is the result of adding $(4 + 7)$ for $4b$ instead of multiplying 4×7.

B is the result of multiplying 4×7 and not adding 13.

D is the result of multiplying 4×8 instead of 4×7, and then adding 13.

18. Correct response: **A**

(*Convert or estimate conversions of measures*)

　　Since 1 liter = 1,000 milliliters (mL), then 2 liters is 2,000 mL. To find how much juice is left, you must subtract the amount that Julie drank: 2,000 mL $-$ 250 mL = 1,750 mL.

Incorrect choices:

B is the result of assuming that 1 liter = 750 mL; 2×750 mL = 1,500 mL; 1,500 mL $-$ 250 mL = 1,250 mL.

C is the result of subtracting what Julie drank (250 mL) from 1 liter (1,000 mL).

D reflects a misunderstanding of the question; 250 mL is the amount of juice that Julie drank.

19. Correct response: **D**

(*Select appropriate unit for measuring length*)

　　A meter is a little longer than a yard, so meters would be the most appropriate unit of measurement.

Incorrect choices:

A A centimeter is less than half an inch long; this unit of measure is too small for measuring a football field.

B A kilometer is about six-tenths of a mile; this unit is too large for measuring the length of a football field.

C A millimeter is one-tenth of a centimeter; this unit of measure is too small for measuring a football field.

20. Correct response: **D**

(*Estimate and find length and area*)

The area of the rectangular yard can be found by multiplying the length × width: 16 ft × 10 ft = 160 square feet. A yard that has double the area would have 2 × 160 square feet, or 320 square feet. If the new yard measures 10 ft × 32 ft, it would have an area of 320 square feet.

Incorrect choices:

A is the result of dividing the original area by 2 instead of multiplying; a yard that measures 8 ft × 10 ft would have one-half the area of the original yard.

B doubles the length to 20 feet, but the width has been divided by 2; the result would be the same as the original area (160 square feet).

C doubles both the length and width of the yard, so the area becomes 640 square feet—four times the original area.

21. Correct response: **C**

(*Estimate and find length, perimeter, and area*)

The perimeter is the distance around the outside of the pool, and the pool is rectangular, so it has two pairs of equal sides. The length of the pool is 12 feet, so two of the sides measure 12 feet: 12 ft × 2 = 24 ft. Subtract 24 from the perimeter (40) to find the length of the other two sides: 40 − 24 = 16 ft. Divide by 2 to find the length of one wall: 16 ÷ 2 = 8 ft. The pool measures 8 ft × 12 ft, so you can find the area by multiplying: 8 ft × 12 ft = 96 sq. ft.

Incorrect choices:

A is the width of the pool.

B is the sum of the length and the width: 8 + 12.

D is the result of multiplying 8 ft × 12 ft and then doubling it to find the area: 2 × 96 = 192.

22. Correct response: **C**

(*Identify and describe the radius and diameter of a circle*)

A line that extends from the center of a circle to any point on the circle is a radius.

22. (continued)

Incorrect choices:

A A chord (such as line segment *AB*) connects two points on the circle.

B The diameter is *AC*; it connects two points on the circle and passes through the center.

D A ray has one endpoint and extends forever in the other direction; there are no rays in this diagram.

23. Correct response: **B**

(*Classify acute, obtuse, and right angles and triangles*)

The triangle Jerry drew has three unequal sides, so it must be a scalene triangle.

Incorrect choices:

A An isosceles triangle has two equal sides.

C An equilateral triangle has three equal sides.

D Jerry's triangle does not have a right angle.

24. Correct response: **D**

(*Identify, classify, and describe plane figures and their attributes*)

The figure shown is a hexagon; it has six sides and six angles.

Incorrect choices:

A A rectangle has four sides and four right angles.

B A pentagon has five sides and five angles.

C An octagon has eight sides and eight angles.

25. Correct response: **C**

(*Determine congruence and similarity*)

Congruent figures have the same shape and the same size. These two rectangles are congruent.

Incorrect choices:

A The rectangles are similar because they have the same shape, but they are not the same size.

B The squares are similar because they have the same shape, but they are not the same size.

D The triangles are neither similar nor congruent.

26. Correct response: **A**

(*Identify, classify, and describe solid figures and their attributes*)

This figure is a rectangular prism; it has three pairs of faces that are congruent and parallel.

Incorrect choices:

B A pyramid is a solid figure with a polygon base and triangular sides.

C A cylinder has two round bases.

D A cone is a solid, pointed figure with one flat, round base.

27. Correct response: **B**

(*Locate and name points on the coordinate plane using ordered pairs*)

In an ordered pair, the first number is the *x*-coordinate, going across to the left or the right. The second number is the *y*-coordinate, going up or down. On the coordinate plane, the trail bars are located at (3, 6).

Incorrect choices:

A (2, 2) is the location of the bottled water.

C (5, 5) is the location of the apples.

D (6, 4) is the location of the street map.

28. Correct response: **D**

(*Locate and name points on the coordinate plane using ordered pairs*)

In an ordered pair, the first number is the *x*-coordinate, going across to the left or the right. The second number is the *y*-coordinate, going up or down. The point described by the ordered pair (2, 5) is 2 across and 5 up; that is the location of the monkeys.

Incorrect choices:

A The polar bears are located at (2, 1).

B The parrots are located at (7, 3).

C The llamas are located at (5, 2).

29. Correct response: **C**

(*Identify and describe points, lines, line segments, and rays*)

This figure is a ray because it has one endpoint and extends forever in one direction.

29. (continued)

Incorrect choices:

A A point names an exact location.

B A line segment has two endpoints.

D A line continues without end in both directions.

30. Correct response: **A**

(*Identify lines of symmetry*)

A figure has a line of symmetry when it can be folded or divided into two parts that are congruent. An equilateral triangle has exactly three lines of symmetry (from each angle to the midpoint of the opposite side).

Incorrect choices:

B This hexagon has two lines of symmetry (one vertical and one horizontal).

C A pentagon has one line of symmetry (vertical).

D A square has four lines of symmetry (one vertical, one horizontal, and two diagonal ones).

31. Correct response: **C**

(*Transform figures in the coordinate plane*)

To determine transformations, visualize the movement of the figures on a coordinate plane to tell whether they slide, flip, or turn. This figure has flipped (over a line) and now is a mirror image of the original figure.

Incorrect choices:

A represents a rotation; the figure has turned 90 degrees.

B is also a rotation; the figure has turned 180 degrees.

D is a translation; the figure has moved position but has not flipped or turned.

32. Correct response: **D**

(*Solve problems involving temperature*)

The temperature dropped 26 degrees (from 26°F to zero) and then another 4 degrees (to -4°F): $26 + 4 = 30$ degrees.

32. (continued)

Incorrect choices:

A is the result of subtracting 4 from 26: $26 - 4 = 22$.

B reflects an error in computation.

C reflects the decrease in temperature from 26°F to zero, but does not include the drop to −4°F.

33. Correct response: **B**

(*Solve problems involving weight/mass*)

One pound = 16 ounces, so the weight of the ostrich egg is 4×16 ounces = 64 ounces.

Incorrect choices:

A is the result of assuming that one pound = 24 ounces: $4 \times 24 = 96$.

C is the result of assuming that one pound = 10 ounces: $4 \times 10 = 40$.

D is the result of assuming that one pound = 8 ounces: $4 \times 8 = 32$.

34. Correct response: **C**

(*Solve problems involving proportions*)

To find the length (n) of the model, you can set up a proportion: $\frac{1 \text{ in.}}{200 \text{ ft.}} = \frac{n}{1,132 \text{ ft.}}$.

Then solve the proportion by multiplying cross products: $200n = 1,132$ ft. To isolate the n on one side of the equation, divide each side by 200: $n = 1,132 \div 200$; $n = 5.66$ inches, or about 6 inches. The model should be about 6 inches long.

Incorrect choices:

A and **B** reflect an incorrect proportion or a misunderstanding of how to use proportions.

D is the result of setting up the proportion correctly, but the result was not rounded correctly.

35. Correct response: **D**

(*Interpret data presented in a line graph*)

The line graph shows average temperatures in January, April, July, and October. The temperature for January is about 58°F, and the temperature for October is about 68°F. To find the difference in temperatures, subtract: $68°F - 58°F = 10°F$.

35. (continued)

Incorrect choices:

A is the result of comparing the temperatures for July and October ($71°F - 68°F = 3°F$).

B is the result of comparing the temperatures for January and April ($63°F - 58°F = 5°F$).

C is the result of comparing the temperatures for July and April ($71°F - 63°F = 8°F$).

36. Correct response: **D**

(*Interpret data presented in a bar graph*)

The graph shows the populations of five countries. According to the graph, the population of Germany is about 82 million. One-half of that number would be about 41 million, which is the approximate population of Spain.

Incorrect choices:

A, **B** and **C** reflects an error in reading the bar graph or in calculating $\frac{1}{2}$ of Germany's population.

37. Correct response: **C**

(*Interpret data in stem-and-leaf plots*)

The number of bowlers who are "20 years old and older" includes all of the bowlers whose ages are listed in Stems 2, 3, 4, and 5. (The second row, for example, represents bowlers aged 20, 21, and 29.) That is a total of 17 bowlers.

Incorrect choices:

A reflects the number of bowlers who are 30 years and older.

B reflects the number of bowlers who are older than 20, but it does not include the one bowler who is exactly 20.

D reflects the total number of bowlers listed in the stem-and-leaf plot, from age 10 to age 52.

38. Correct response: **B**

(*List all possible outcomes or construct sample spaces using lists, charts, frequency tables, and tree diagrams*)

   In one roll of the number cube, there are 6 possible outcomes: 1, 2, 3, 4, 5, and 6. The 2 and 3 represent $\frac{1}{3}$ of the outcomes (2 out of 6). In 12 rolls, Latoya is likely to roll a 2 or a 3 $\frac{1}{3}$ of the time, or 4 times.

Incorrect choices:

A is the likelihood of rolling one of the numbers. For example, the likelihood of rolling a 2 is 1 out of 6; in 12 rolls, the likelihood is $\frac{1}{6}$ of 12, or 2 times.

C is the likelihood of rolling any of three numbers (6 times out of 12).

D is the likelihood of rolling any of four numbers, or any number *other than* a 2 or a 3 (8 times out of 12).

39. Correct response: **B**

(*Find probabilities, represented as ratios, decimals, or percents*)

   In one spin, there are four possible outcomes: red, green, yellow, or blue. The probability of getting any one of the colors is $\frac{1}{4}$, so you can expect to get yellow about 1 in every 4 tries. In 40 spins, you can expect the spinner to land on yellow about 10 times ($\frac{1}{4}$ of 40).

Incorrect choices:

A reflects a misunderstanding of probability: four is the number of outcomes for any one spin.

C reflects a misunderstanding of probability; 15 is the number of times the spinner would likely land on yellow in 60 tries.

D reflects the number of times the spinner would land on yellow and one other color (about $\frac{1}{2}$ the time).

40. Correct response: **A**

(*Determine and compare probabilities for simple and compound events*)

   There are nine letter tiles in the bag, and three of them are vowels (A, E, I). The probability of picking a vowel in one try is 3 of 9, or $\frac{1}{3}$.

Incorrect choices:

B is the result of comparing the number of vowels to the number of consonants (3 of 6, or $\frac{1}{2}$) instead of the total number of letter tiles (9).

C is the probability of getting a consonant (6 of 9, or $\frac{2}{3}$).

D is the result of counting only two vowels in the letter tiles instead of three vowels.

41. Correct response: **B**

(*Determine and describe the mean, median, mode, and range of data*)

   To find the median, list the numbers from the table in numerical order: 23, 37, 70, 75, 81, 89, 139. The median is the number in the middle of the set of numbers: 75.

Incorrect choices:

A reflects an error in finding the middle number; 81 is the fourth number in the set.

C is the approximate mean of the data (rounded to the nearest whole number), and it is about halfway between the third and fourth numbers (70 and 75).

D reflects an error in finding the middle number; 70 is the third number of seven.

42. Correct response: **C**

(*Determine and describe the mean, median, mode, and range of data*)

   To find the range of the numbers in this table, subtract the lowest number (23) from the highest number (139): $139 - 23 = 116$.

42. (continued)
   Incorrect choices:

   A is the result of subtracting incorrect numbers (for example, $139 - 89$).

   B is the result of subtracting the second-lowest number (37) from 139.

   D is the result of adding the highest and lowest numbers ($139 + 23$) instead of subtracting.

43. Correct response: **B**
   (*Collect, organize, display, and interpret data to solve problems*)

   The circle graph (or pie chart) shows the percentage of each kind of book Beth read. It shows that 45% of the books she read were mysteries. You can find the number of mystery books she read by converting the percentage to a decimal and multiplying it by the total number of books she read (20): $0.45 \times 20 = 9$.

   Incorrect choices:

   A is the number of science-fiction books she read: 25% of 20 books.

   C reflects an error in reading the graph, seeing mystery books as 50% of the total, or in computing the number of mystery books.

   D is the percent of mystery books, not the actual number.

44. Correct response: **D**
   (*Identify, describe, and extend numerical patterns*)

   To determine the number of stamps Jack will have in January, figure out the pattern and then extend it through January. The pattern shows that Jack adds 12 stamps to his collection every two months (for example, $19 + 12 = 31$), so the pattern is "+ 12" (or "+ 6" every month). In January, Jack will have $55 + 12 + 12 + 12 = 91$ stamps.

44. (continued)
   Incorrect choices:

   A is the number of stamps Jack will have in November ($55 + 12 + 12$).

   B reflects an error in figuring the number of months: 79 stamps in November + 6 stamps for the next month = 85 stamps.

   C reflects a mathematical error in calculating the number of stamps.

45. Correct response: **A**
   (*Interpret, write, and simplify algebraic expressions*)

   If $a = 4$, then $2a = 2 \times 4$, or 8. The expression can be written as $248 \div 8$, and the quotient is 31.

   Incorrect choices:

   B is the result of dividing 248 by 4, the value of a, instead of 8, the value of $2a$.

   C is the result of dividing 248 by 2, instead of $2a$, and omitting the value of a.

   D is the result of adding $248 + 8$ instead of dividing.

46. Correct response: **C**
   (*Apply basic properties and order of operations with algebraic expressions*)

   The school bought 20 tickets for students at $18 a piece ($20 \times \$18$) and 4 tickets for adults at $30 ($4 \times \$30$). The sum of these two purchases can be expressed as (20×18) + (4×30).

   Incorrect choices:

   A The expression adds the cost of the student ticket and the cost of the adult ticket ($18 + 30$) and multiplies by the total number of tickets (24).

   B The expression multiplies the two costs instead of adding.

   D The expression subtracts the cost of the adult tickets from the cost of the student tickets instead of adding the two costs.

47. Correct response: **D**
(*Use simple equations to represent problem situations*)

The number sentence has a variable (x), meaning that one of the numbers is unknown. Jack drove 420 miles in 7 hours, so x represents his average driving speed or average number of miles he drove per hour. You can describe this situation as 7 times the number of miles per hour, or $7x = 420$ miles.

Incorrect choices:

A The unknown is the number of books June started with; the situation can be represented as $x - 7 = 420$.

B The unknown is the total number of hours Theo worked; the situation can be represented as $7 + 7 + 7 = x$.

C The unknown is the number of beads each friend received; this situation can be expressed as $420 \div 7 = x$.

48. Correct response: **B**
(*Represent and describe mathematical relationships with graphs*)

The graph shows the relationship of Mariel's different expenses to the total amount she spent. Since the hotel cost was 50% of her expenses, you can determine that she spent one-half of her money on the hotel.

Incorrect choices:

A Meals and activities made up 40% of her costs (25% + 15%), which was not "most of her money."

C Hotel and activities made up 65% of her costs (50% + 15%), while meals only represented 25%.

D Meals and souvenirs made up 35% of her costs (25% + 10%), and 35% is more than one-quarter, not less.

49. Correct response: **B**
(*Solve one-step equations*)

Write an equivalent equation with multiplication: $8x = 72$. Find the number that when multiplied by 8 results in 72.

Incorrect choices:

A reflects an error in division or in factoring, thinking that $72 \div 8 = 7$.

C reflects an error in division, thinking that $72 \div 8 = 12$.

D is the result of adding $72 + 8 = 80$.

50. Correct response: **D**
(*Make predictions from patterns, data, or examples*)

To find the 15th figure, you must first determine the pattern in the sequence of figures and then extend it to 15 figures. The pattern begins with four white boxes in the square and then has one black box in each square; the location of the black box starts in the upper left and moves around the square in clockwise rotation. The pattern consists of five figures, so the same figure will appear every five times: 5, 10, 15, and so on. The 15th figure will be the same as the 5th.

Incorrect choices:

A is the first figure in the pattern; it will also be the 6th, the 11th, and the 16th.

B is the third figure in the pattern; it will also be the 8th, the 13th, and the 18th.

C is the second figure in the pattern; it will also be the 7th, the 12th, and the 17th.

1. C	**11.** C	**21.** C	**31.** C	**41.** C
2. A	**12.** D	**22.** B	**32.** B	**42.** B
3. D	**13.** B	**23.** C	**33.** B	**43.** A
4. D	**14.** D	**24.** A	**34.** A	**44.** B
5. B	**15.** C	**25.** D	**35.** D	**45.** D
6. A	**16.** B	**26.** B	**36.** A	**46.** C
7. C	**17.** A	**27.** C	**37.** C	**47.** A
8. B	**18.** C	**28.** B	**38.** B	**48.** B
9. B	**19.** D	**29.** A	**39.** C	**49.** B
10. C	**20.** B	**30.** C	**40.** B	**50.** D

Answer Key Explanations

1. Correct response: **C**
(*Compare and order decimals*)

The number line extends from 8.4 to 8.5 and is divided in ten parts: 8.41, 8.42, and so on. Each hash mark on the number line represents 0.01. To find a number that is "0.03 greater than 8.43," first find 8.43 on the number line and then add 0.03 by moving three hash marks to the right. The result is 8.46.

Incorrect choices:

A Point *A* represents 8.4, or 8.40, the first number on the number line.

B Point *B* represents 8.43.

D Point *D* represents 8.49; that is 0.06 greater than 8.43.

2. Correct response: **A**
(*Compare, order, and use percents*)

The Rovers won 12 games and lost 4 games, so they played a total of 16 games (12 + 4). They lost 4 of 16 games, or $\frac{4}{16}$. To find the percentage, divide $\frac{4}{16} = \frac{1}{4} = 25\%$.

2. (continued)
Incorrect choices:

B is the result of comparing 4 games to 12 games (the number of games won); $\frac{4}{12} = \frac{1}{3} = 33\frac{1}{3}\%$.

C is the result of comparing the difference between games won (12) and games lost (4) to the number of games won: $\frac{8}{12} = \frac{2}{3}$, which is about 67%.

D is the result of comparing the number of games won (12) to the total number of games: $\frac{12}{16} = \frac{3}{4} = 75\%$.

3. Correct response: **D**
(*Compare, order, and use integers*)

Sea level is considered 0 elevation, so the ship dropped its anchor from 5 feet above 0 to a depth of 22 feet below 0: 5 + 22 = 27 feet. You can also use the absolute value of these two numbers (5 + 22) to reach the same answer.

3. (continued)
Incorrect choices:

A is the result of a mathematical error in subtracting.

B is the result of subtracting 22 ft − 5 ft = 17 ft.

C is the result of a mathematical error in adding 5 and −22.

4. Correct response: **D**
(*Identify and use place value*)
   This number can be expressed as $700 + 30 + 2 + 0.5 + 0.06 + 0.00\mathbf{8}$. Digits to the right of the decimal point represent tenths, hundredths, and thousandths, in that order. The **8** represents the thousandths place.

Incorrect choices:

A The 5 represents the tenths place.

B The 6 represents the hundredths place.

C The 7 represents the hundreds place.

5. Correct response: **B**
(*Identify factors and multiples*)
   Mark wants to put the same number of books in each box, so the number of books per box must be a factor of 84 (or, the number 84 must be divisible by the number of books per box, without any books left over). Mark can put 14 books in each box because 14 is a factor of 84: $7 \times 14 = 84$.

Incorrect choices:

A 16 is not a factor of 84: $84 \div 16 = 5$ R4.

C 10 is not a factor of 84: $84 \div 10 = 8$ R4.

D 8 is not a factor of 84: $84 \div 8 = 10$ R4.

6. Correct response: **A**
(*Use expanded notation and exponents*)
   The number 5^3 is the same as $5 \times 5 \times 5$, or 125.

Incorrect choices:

B is the result of multiplying $5 \times 3 \times 3$.

C is the result of multiplying 5×5, or 5^2.

D is the result of multiplying 5 by the exponent 3: $5 \times 3 = 15$.

7. Correct response: **C**
(*Use ratios to describe and compare two sets of data*)
   The Student Council has a total of 12 members (5 boys + 7 girls). The ratio of the number of girls to the total number of members is the ratio of part (7) to the whole (12), or 7 to 12.

Incorrect choices:

A is the result of comparing the number of boys (5) to the number of girls (7).

B is the result of comparing the number of boys (5) to the total number of members (12).

D is the result of comparing the whole to the part, or the total number of members (12) to the number of girls (7).

8. Correct response: **B**
(*Add, subtract, multiply, and divide whole numbers*)
   To find the difference in area between Hawaii and New Hampshire, subtract: 10,931 sq mi − 9,350 sq mi = 1,581 sq mi.

Incorrect choices:

A is the result of a mathematical error in subtracting (forgetting to subtract the thousands).

C is the result of subtracting incorrectly (subtracting 3 from 5 in the tens column instead of 5 from 3).

D is the result of adding the two areas (10,931 + 9,350) instead of subtracting.

9. Correct response: **B**
(*Add, subtract, multiply, and divide fractions*)
   Darrell buys $\frac{3}{4}$ pound of potato salad and gives away $\frac{2}{5}$ of that amount. To find how much he gives away, multiply $\frac{2}{5}$ times $\frac{3}{4}$ lb: $\frac{2}{5} \times \frac{3}{4} = \frac{6}{20}$, or $\frac{3}{10}$ lb.

9. (continued)
   Incorrect choices:

   A is the result of adding the numerators and multiplying the denominators: $\frac{5}{20}$, or $\frac{1}{4}$.

   C is the result of subtracting $(\frac{15}{20} - \frac{8}{20})$ instead of multiplying.

   D is the result of multiplying the numerators and adding the denominators.

10. Correct response: **C**
   (*Add, subtract, multiply, and divide decimals*)
   Camille rides 2.7 miles to school and 2.7 miles home each day, so the total number of miles per day is $2.7 + 2.7 = 5.4$ miles. In 5 days, she rides 5 times that amount: $5 \times 5.4 = 27$ miles.

   Incorrect choices:

   A is the result of adding 2.7 miles + 2.7 miles + 5 days.

   B is the result of including only one of Camille's trips each day, multiplying 5×2.7 miles = 13.5 miles.

   D is the result of multiplying 2.7×2.7 instead of adding the two distances, and then multiplying times 5, the number of days.

11. Correct response: **C**
   (*Solve problems involving integers*)
   The Rams gained 23 yards ($+23$), lost 7 yards (-7), and gained 5 yards ($+5$). The total number of yards they gained or lost can be expressed as $23 - 7 + 5$; $23 - 7 = 16$, and $16 + 5 = 21$ yards.

   Incorrect choices:

   A is the result of subtracting all three numbers: $-23 - 7 - 5$.

   B is the result of subtracting $23 - 7 = 16$, and then subtracting 5 yards instead of adding: $16 - 5 = 11$.

   D is the result of adding all three numbers: $23 + 7 + 5$.

12. Correct response: **D**
   (*Estimate and round using whole numbers and fractions*)
   To estimate the number of miles Ben walked, round each mixed number to the nearest whole number: $2 + 3 + 4 = 9$ miles. This is the number of miles he has walked so far. To find the number he must walk to reach his goal, subtract the number he has walked from the total number of miles: $15 - 9 = 6$ miles.

   Incorrect choices:

   A represents an estimate of the miles he has walked so far.

   B is the result of rounding all three mixed numbers to the lowest whole number: $2 + 3 + 3 = 8$ miles.

   C is the result of rounding all three numbers to the lowest whole number ($2 + 3 + 3 = 8$ miles) and then subtracting: $15 - 8 = 7$ miles.

13. Correct response: **B**
   (*Use order of operations to simplify whole-number expressions*)
   To simplify, complete operations in parentheses first, then simplify exponents, and then add and subtract in order from left to right:

   $$(21 - 6) + 3^2 - (8 \times 2) = 15 + 3^2 - 16$$
   $$= 15 + 9 - 16 = 8$$

   Incorrect choices:

   A is the result of an error in simplifying 3^2 to 6 instead of 9: $15 + 6 - 16 = 5$.

   C and **D** are the results of an error in the order of operations.

14. Correct response: **D**
   (*Apply the properties of operations*)
   The inverse of multiplication is division; $30 \times 20 = 600$. So the inverse operation is $600 \div 30 = 20$.

14. (continued)
   Incorrect choices:

   A $30 + 20$ is the inverse of subtraction, not multiplication.

   B $30 - 20$ is the inverse of addition, not multiplication.

   C The inverse of multiplication is division, but $30 \div 20$ uses incorrect numbers.

15. Correct response: **C**
   (*Solve multi-step problems involving whole numbers*)
   The students washed 24 cars at $7 apiece ($24 \times \$7$) and 15 vans at $9 apiece ($15 \times \$9$). $24 \times \$7 = \$168$, and $15 \times \$9 = \$135$. The total they earned is the sum of $\$168 + \$135$, which is $303.

   Incorrect choices:

   A represents the money collected for washing 15 vans ($\$15 \times 9 = \$135$).

   B represents the money collected for washing 24 cars ($24 \times \$7 = \$168$).

   D represents an estimated number of cars and vans washed ($24 + 15 = 20 + 20$, or 40) and the estimated amount ($8); 40 cars and vans $\times \$8 = \$320$.

16. Correct response: **B**
   (*Solve multi-step problems involving whole numbers and fractions*)
   Of the 24 students, $\frac{1}{2}$ voted for the aquarium: $\frac{1}{2} \times 24 = 12$. And, $\frac{1}{8}$ of the students voted for the science museum: $\frac{1}{8} \times 24 = 3$. The total number of students choosing the aquarium (12) or the science museum (3) was $12 + 3 = 15$.

   Incorrect choices:

   A represents the number of students voting for the aquarium (12) and the planetarium ($\frac{1}{4} \times 24 = 6$).

   C represents the number of students voting for the aquarium only (12).

   D represents the number of students voting for the planetarium (6) and either the science museum or the shoe factory (3).

17. Correct response: **A**
   (*Solve number sentences with one variable*)
   If $y = 8$, then $2y$ can be expressed as $2(8)$, or 16. The expression becomes $(56 - 16) \div 4$, or $40 \div 4 = 10$.

   Incorrect choices:

   B represents the value of $2y$ ($2 \times 8 = 16$).

   C is the result of subtracting $56 - 16$ but not dividing by 4.

   D is the result of adding $56 + 16$ and not dividing by 4.

18. Correct response: **C**
   (*Convert or estimate conversions of measures*)
   A kilogram is about 2.2 pounds, so a dog that weighs 9 kilograms weighs about 20 pounds.

   Incorrect choices:

   A is the result of dividing 9 kilograms by 2.2 instead of multiplying, and then rounding the result to 4.

   B is the dog's weight in kilograms, rounded to the nearest 10; or it could reflect an error in thinking that a kilogram is almost the same as a pound.

   D is the result of confusing pounds and ounces, thinking that one kilogram equals 16 pounds ($9 \times 16 = 144$).

19. Correct response: **D**
   (*Select appropriate unit for measuring capacity*)
   The capacity of a swimming pool would best be measured in gallons, which is the largest unit of capacity offered as an answer choice.

   Incorrect choices:

   A A ton is a unit of weight, not capacity.

   B Ounces are too small to use for measuring the capacity of a swimming pool.

   C Quarts are smaller than gallons and would be less appropriate for measuring the capacity of a swimming pool.

20. Correct response: **B**

(*Estimate and find length and area*)

The kite is triangular, and the area of a triangle can be determined with the formula $A = \frac{1}{2}bh$. The base is 20 inches, and the height is 12 inches. $A = \frac{1}{2}$ (20 in. $\times$ 12 in.); 240 sq. in. $\div$ 2 = 120 sq. in.

Incorrect choices:

A is the result of multiplying 20 in. $\times$ 12 in. and dividing by 3 instead of 2.

C is the result of determining $\frac{2}{3}$ of 240 square inches instead of $\frac{1}{2}$.

D is the result of multiplying the base times the height (20 in. $\times$ 12 in.) but not dividing by 2.

21. Correct response: **C**

(*Estimate and find length and area*)

The area of the garden can be determined by dividing the garden into two rectangles and then finding the area of each rectangle. The larger rectangle would be 6 m $\times$ 4 m = 24 m². The smaller rectangle would be 3 m $\times$ 3 m = 9 m². The area of the garden is the sum of these two figures: 24 m² + 9 m² = 33 m².

Incorrect choices:

A is the area of the larger rectangular portion of the garden (4 m $\times$ 6 m).

B is the perimeter of the garden, the result of adding all the dimensions given in the figure: 6 m + 7 m + 4 m + 3 m + 3 m + 3 m = 26 m.

D is the area of the garden as if it were a complete rectangle measuring 6 m $\times$ 7 m.

22. Correct response: **B**

(*Identify and describe the radius, diameter, and circumference of a circle*)

The circumference of a circle can be determined by the formula $C = \pi d$, where d is the diameter (8 ft). For the round table, the circumference is $\pi \times$ 8 ft.

22. (continued)

Incorrect choices:

A is the result of using an incorrect formula, $C = \pi r$, where r is the radius of the circle.

C is the area of the circle, based on the formula $A = \pi r^2$.

D is the result of using an incorrect formula for area: $A = \pi d^2$.

23. Correct response: **C**

(*Classify acute, obtuse, and right angles and triangles*)

The triangle shown is an acute triangle because it has three acute angles (each measuring less than 90°).

Incorrect choices:

A This figure is an isosceles right triangle.

B This triangle is an obtuse triangle because it has an obtuse angle (greater than 90 degrees).

D This triangle is an obtuse triangle because it has an obtuse angle (greater than 90 degrees).

24. Correct response: **A**

(*Identify, classify, and describe plane figures and their attributes*)

A rhombus is a quadrilateral because it has four sides; all four sides are equal, and the pairs of opposite sides are parallel.

Incorrect choices:

B describes a triangle.

C describes a trapezoid.

D describes a hexagon.

25. Correct response: **D**

(*Determine congruence and similarity*)

Similar figures have the same shape but not the same size; congruent figures are the same shape and the same size. These rectangles are similar and congruent.

25. (continued)

Incorrect choices:

A The rectangle and square are not similar or congruent.

B These rectangles are similar, but they are not the same size so they are not congruent.

C These squares are similar, but they are not the same size so they are not congruent.

26. Correct response: **B**
(*Identify, classify, and describe solid figures and their attributes*)

This figure is a cone because it is a pointed figure with a flat, round base.

Incorrect choices:

A A cylinder has two flat, round bases.

C A triangle is a plane figure with three sides.

D A pyramid has a square or triangular base and triangular sides.

27. Correct response: **C**
(*Locate and name points on coordinate plane using ordered pairs*)

In an ordered pair, the first number is the *x*-coordinate, going across to the left or the right. The second number is the *y*-coordinate, going up or down. The cashier is located at (2, 6). To find the location of the salad bar, you would go 4 units to the right from the cashier and 3 units down to (6, 3).

Incorrect choices:

A The location (4, 3) is 2 units across and 3 units down from the cashier.

B The location (4, 6) is 2 units across from the cashier.

D The location (5, 2) is 3 units across and 4 units down from the cashier.

28. Correct response: **B**
(*Locate and name points on coordinate plane using ordered pairs*)

In an ordered pair, the first number is the *x*-coordinate, going across to the left or the right. The second number is the *y*-coordinate, going up or down. Point *G* is located at (5, 5).

28. (continued)

Incorrect choices:

A Point *F* is located at (2, 5).

C Point *H* is located at (1, 3).

D Point *J* is located at (4, 3).

29. Correct response: **A**
(*Identify and describe points, lines, line segments, and rays*)

The figure *MN* is a line because it extends forever in both directions.

Incorrect choices:

B Figure *PQ* is a line segment because it has two endpoints.

C *R* describes a point because it names one exact location.

D Figure *ST* is a ray because it has one endpoint and extends forever in the other direction.

30. Correct response: **C**
(*Identify lines of symmetry*)

A figure has line symmetry when it can be folded or divided into two parts that are congruent. A square has four lines of symmetry (one vertical, one horizontal, and two diagonal ones).

Incorrect choices:

A A regular pentagon has five lines of symmetry.

B This trapezoid has only one line of symmetry.

D An equilateral triangle has three lines of symmetry.

31. Correct response: **C**
(*Transform figures in the coordinate plane*)

To determine transformations, look at the movement of the figures to tell whether they slide, flip, or turn. In this pair, the second figure shows a translation because it has slid to a new location but has not flipped or turned.

31. (continued)
Incorrect choices:

A The second figure shows a flip, or reflection, across the dotted line.

B The second figure shows a turn, or rotation, of 90°.

D The second figure shows a turn, or rotation, of 180°.

32. Correct response: **B**
(*Solve problems involving length and area*)
The area of a rectangle can be determined by multiplying length × width. To find the length of this shed when you know the area (96 sq ft) and the width (8 ft), you can divide the area by the width: 96 sq ft ÷ 8 ft = 12 feet.

Incorrect choices:

A is the width of the rectangle, not the length.

C is the result of adding the length and the width (8 ft + 12 ft).

D is the perimeter of the shed: (2 × 8 ft) + (2 × 12 ft).

33. Correct response: **B**
(*Solve problems involving length and volume*)
The volume of a rectangular box can be determined by multiplying length × width × height: 2 in. × 5 in. × 12 in. = 120 cubic inches.

Incorrect choices:

A is the result of multiplying 2 in. × 5 in. × 12 in, and then doubling the product.

C is the result of multiplying 2 in. × 12 in. and 5 in. × 12 in., and then adding the products: 24 + 60 = 84.

D is the area of one side, the result of multiplying length × width (5 in. × 12 in.).

34. Correct response: **A**
(*Solve problems involving proportions*)
To find the actual distance from Bell Harbor to Wayland, you can set up a proportion and multiply cross products: $\dfrac{1 \text{ in.}}{4.5 \text{ miles}} = \dfrac{4 \text{ in.}}{x}$.

$1x = 4 \times 4.5$ miles, or $x = 18$ miles

34. (continued)
Incorrect choices:

B is the result of rounding 4.5 down to 4 and multiplying 4 × 4.

C is the result of subtracting 4.5 miles from the total of 18 miles.

D is the result of adding 4 (inches) + 4.5 miles = 8.5 miles.

35. Correct response: **D**
(*Interpret data presented in a chart*)
To find the difference between the subway systems in New York City and Tokyo, subtract: 231 miles − 160 miles = 71 miles.

Incorrect choices:

A is the result of adding the length of New York City's subway system and the length of the Tokyo system: 231 miles + 160 miles.

B is the difference between the systems in London and Tokyo: 244 miles − 160 miles.

C is the difference between the systems in London and Moscow: 244 miles − 163 miles.

36. Correct response: **A**
(*Interpret data presented in a line graph*)
To find the lowest average temperature, compare the data points on the line graph. The lowest point is the first on the graph, the average temperature for the month of January.

Incorrect choices:

B reflects a misreading of the graph; March has the third-lowest average temperature.

C reflects a misunderstanding of the question; July has the highest average temperature.

D reflects a misreading of the graph; November has the second-lowest temperature.

37. Correct response: **C**
(*Interpret data in stem-and-leaf plots*)
The number of plants that are taller than 29 inches are the plants that are listed at 30 inches and higher; these include all the values for stems 3 and 4, a total of 9 plants.

37. (continued)
Incorrect choices:

A represents the number of plants that are 29 inches or shorter.

B represents the number of plants that are 30 to 36 inches tall.

D represents the total number of plants listed on the plot.

38. Correct response: **B**
(*List all possible outcomes or construct sample spaces using lists, charts, frequency tables, and tree diagrams*)

The breakfast special offers three choices of main dish, three choices for the side dish, and two choices for the drink, so the total number of combinations of one from each group is $3 \times 3 \times 2 = 18$. Choice B is the only correct list that includes all 18 combinations.

Incorrect choices:

A has only three possible combinations; it lists the choices in each row of the chart from left to right.

C has nine combinations; it is incorrect because it lists two drinks with each combination of main dish and side dish.

D lists the choices from each column in sequence and does not include any correct combinations of one choice from each group.

39. Correct response: **C**
(*Determine and compare probabilities for simple and compound events*)

To find the most likely outcome, compare the probabilities of all five colors: 3 of the 60 ducks are red, so the probability of choosing red is $\frac{3}{60}$; orange is $\frac{10}{60}$; yellow is $\frac{20}{60}$; blue is $\frac{12}{60}$; and purple is $\frac{15}{60}$. The color with the highest probability is yellow.

Incorrect choices:

A Red has the lowest probability and is least likely to be chosen.

B and **D** The probability of getting an orange or purple duck is lower than the probability of getting yellow.

40. Correct response: **B**
(*Find probabilities represented as decimals*)

There are 12 blue ducks out of 60 in the Duck Pond, so the probability of choosing a blue duck is $\frac{12}{60}$, or $\frac{1}{5}$, which is equivalent to the decimal 0.2.

Incorrect choices:

A is the result of changing the number of blue ducks (12) to a decimal number (0.12), or determining that the probability of choosing a blue duck is $\frac{12}{100}$.

C is the probability of choosing a purple duck ($\frac{15}{60} = \frac{1}{4}$, or 0.25).

D is the approximate probability of choosing a yellow duck ($\frac{20}{60} = \frac{1}{3}$, or about 0.33).

41. Correct response: **C**
(*Collect, organize, display, and interpret data to solve problems*)

A circle graph would be best for these data because they represent parts of a whole and can be shown as parts of a circle.

Incorrect choices:

A A line graph is most appropriate for showing change over a period of time.

B A line plot is most appropriate for showing the frequency of different data values.

D A stem-and-leaf plot is most appropriate for showing frequencies.

42. Correct response: **B**
(*Determine and describe the mean, median, mode, and range of data*)

To find the median number of visitors, list the data from the chart in order from least to greatest: 2.7, 2.9, 3.2, 4.4, 9.3. The middle number in the list (3.2) is the median.

Incorrect choices:

A represents the average of 2.7 and 2.9, the first and second numbers in the list.

C represents the fourth number in the list.

D is the range of the data ($9.3 - 2.7$).

43. Correct response: **A**

(*Determine and describe the mean, median, mode, and range of data*)

The mean is the sum of the numbers of visitors divided by the number of parks: $4.4 + 9.3 + 2.7 + 2.9 + 3.2 = 22.5 \div 5 = 4.5$.

Incorrect choices:

B is the result of adding the five numbers and dividing by 4 instead of 5.

C is the range of the data $(9.3 - 2.7)$.

D is the sum of the five numbers in the chart.

44. Correct response: **B**

(*Identify, describe, and extend numerical patterns*)

To predict the next three numbers, you must first find the rule for the pattern in the y column and then extend the pattern. The rule for the pattern is "$+ 3$," so the next three numbers will be 10, 12, and 14.

Incorrect choices:

A is the result of adding 3, then 4, then 5 to the x values.

C is the result of adding 2 instead of 3.

D is the result of adding 1 instead of 3.

45. Correct response: **D**

(*Interpret, write, and simplify algebraic expressions*)

If $x = 5$, then $3x$ is equal to $3(5)$, or 15. The expression can be written as $(15 - 4) - 3$, or $11 - 3 = 8$.

Incorrect choices:

A is the result of multiplying $3(5)$ and then adding 4 and 3 instead of subtracting.

B is the result of substituting 5 for x in $3x$ and then not calculating the rest of the expression.

C is the result of multiplying $3(5)$ and subtracting 4 but not subtracting 3.

46. Correct response: **C**

(*Apply basic properties and order of operations with algebraic expressions*)

Using the distributive property, this expression can be written as $(6 \times 4) + (6 \times 3)$.

Incorrect choices:

A (4×3) is not equivalent to $(4 + 3)$.

B Moving the parentheses changes the value of the expression.

D This expression reverses the operations of addition and multiplication, and these operations are not interchangeable.

47. Correct response: **A**

(*Use simple equations to represent problem situations*)

The number sentence has a variable (x), meaning that one of the numbers is unknown. Katya had 63 CDs and gave an unknown number of them to each of 7 friends, so x represents the number of CDs she gave to each friend. You can describe this situation as $63 \div x = 7$.

Incorrect choices:

B The unknown is the number of books Carl has not read; the situation can be represented as $63 - 7 = x$.

C The unknown is the total number of hours Sam worked; the situation can be represented as $63 + 7 = x$.

D The unknown is the amount of money Tony has in his pocket; this situation can be expressed as $63 \times 7 = x$.

48. Correct response: **B**

(*Represent and describe mathematical relationships with lists, tables, and charts*)

The table shows a pattern in Brenda's bowling scores. The best description of the pattern is that her scores increase by 7 points each game (for example, $89 + 7 = 96$).

48. (continued)

Incorrect choices:

A Brenda's scores did not decrease; they increased.

C Brenda steadily improved her scores, but she did not double her score from Game 1 (89) to Game 5 (117).

D Brenda's scores improved with each game, but the improvement was not more than 10 percent (for example, the improvement from Game 1 to Game 2 was $\frac{7}{89}$, or less than 8%).

49. Correct response: **B**
(*Solve one-step equations*)

To solve for x, you must first isolate x on one side of the equation. You can do this by dividing each side by 3:

$$\frac{3x}{3} = \frac{159}{3}, x = 53$$

Incorrect choices:

A reflects a low estimate, resulting in $\frac{150}{3}$.

C reflects a high estimate, resulting in $\frac{165}{3}$.

D is the result of multiplying 3×159 instead of dividing.

50. Correct response: **D**
(*Make predictions from patterns, data, or examples*)

To find the number of miles for Week 6, you must first determine the pattern in the table and then extend it to Week 6. The pattern for number of miles is "+ 3," so Kevin will probably run 14 miles in Week 5 (11 + 3) and 17 miles in Week 6 (11 + 3 + 3).

Incorrect choices:

A is the result of adding 1 mile to Week 4 (11 + 1).

B is the number of miles Kevin will run in Week 5.

C is the result of adding 2 miles each week instead of 3 miles.

Student Scoring Chart

Student Name _____

Teacher Name _____

Test 1	Item Numbers	No. Correct/ Total	Percent (%)
Number and Number Sense	1–7	/7	
Operations	8–17	/10	
Measurement and Geometry	18–34	/17	
Statistics and Probability	35–43	/9	
Patterns, Relations, and Algebra	44–50	/7	
Total	**1–50**	**/50**	

Test 2	Item Numbers	No. Correct/ Total	Percent (%)
Number and Number Sense	1–7	/7	
Operations	8–17	/10	
Measurement and Geometry	18–34	/17	
Statistics and Probability	35–43	/9	
Patterns, Relations, and Algebra	44–50	/7	
Total	**1–50**	**/50**	

Test 3	Item Numbers	No. Correct/ Total	Percent (%)
Number and Number Sense	1–7	/7	
Operations	8–17	/10	
Measurement and Geometry	18–34	/17	
Statistics and Probability	35–43	/9	
Patterns, Relations, and Algebra	44–50	/7	
Total	**1–50**	**/50**	

Comments/Notes:_____

_____

Standardized Test Tutor: Math Grade

Classroom Scoring Chart

Teacher Name _____

Student Name	Test 1	Test 2	Test 3

Notes:

Notes:

Standardized Test Tutor: Math (Grade 5)